I0034611

Maurizio Dapor
Electron–Atom Collisions

Also of Interest

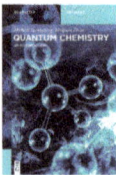

Quantum Chemistry
An Introduction
Springborg, Zhou, 2021
ISBN 978-3-11-074219-0, e-ISBN 978-3-11-074220-6

Ion-Atom Collisions
The Few-Body Problem in Dynamic Systems
Schulz (Ed.), 2019
ISBN 978-3-11-057942-0, e-ISBN 978-3-11-058029-7

Coherent Quantum Physics
A Reinterpretation of the Tradition
Neumaier, 2021
ISBN 978-3-11-066729-5, e-ISBN 978-3-11-066738-7

Electrons in Solids
Mesoscopics, Photonics, Quantum Computing, Correlations, Topology
Bluhm, Brückel, Morgenstern, von Plessen, Stampfer, 2019
ISBN 978-3-11-043831-4, e-ISBN 978-3-11-043832-1

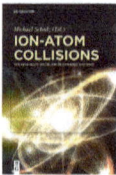

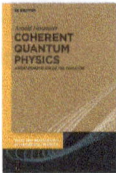

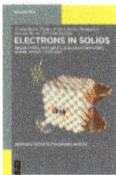

Maurizio Dapor

Electron–Atom Collisions

Quantum-Relativistic Theory and Exercises

DE GRUYTER

Author
Dr. Maurizio Dapor
European Centre for Theoretical Studies in
Nuclear Physics and Related Areas
Fondazione Bruno Kessler
Strada delle Tabarelle 286
38123 Villazzano
Trento
Italy
dapor@ectstar.eu

ISBN 978-3-11-067535-1
e-ISBN (PDF) 978-3-11-067537-5
e-ISBN (EPUB) 978-3-11-067541-2

Library of Congress Control Number: 2021950452

Bibliographic information published by the Deutsche Nationalbibliothek
The Deutsche Nationalbibliothek lists this publication in the Deutsche Nationalbibliografie;
detailed bibliographic data are available on the Internet at http://dnb.dnb.de.

© 2022 Walter de Gruyter GmbH, Berlin/Boston
Cover image: merrymoonmary / E+ / Getty Images
Typesetting: VTeX UAB, Lithuania
Printing and binding: CPI books GmbH, Leck

www.degruyter.com

To my mother and father

Preface

This book deals with collisions of electrons with atoms. Both the nonrelativistic and the relativistic theories are presented here. Since we are interested in applications, the first part of the book is devoted to the basic concepts of computational physics, describing the main numerical tools necessary for solving problems concerning the scattering of charged particles by central fields. We also briefly describe the main special functions of mathematical physics and provide methods to numerically calculate them. The second part of the book is dedicated to the nonrelativistic approach to the study of electron–atom scattering and to an introduction to Pauli matrices and spin. The Thomas Fermi and Hartree–Fock methods for describing many-electron atoms and, in particular, for calculating the so-called screening function are described in the second part of the book. The screening function is crucial for the calculation of phase shifts, and its analytical approximation is also presented to make easier the calculation of the electrostatic atomic potential. In the third part of the volume, after an introduction to the quantum relativistic equations (Klein–Gordon equation and Dirac equation), the Mott theory is described. It represents the quantum-relativistic theory of elastic scattering of electrons by central fields, the so-called relativistic partial wave expansion method. The last part of the book presents several applications. It contains exercises devoted to the calculation of the special functions of mathematical physics (notably, Legendre polynomials and spherical Bessel functions, both regular and irregular) and to their use for computing phase shifts, scattering amplitudes, differential elastic scattering cross-sections, and spin-polarization parameters. The exercises are provided with an increasing degree of difficulty. With the aid of these exercises, the reader can use all the information described in the first three parts of the book to write her/his own computer codes for the computation of all the quantities relevant to the scattering processes.

I am indebted to all my colleagues at the European Centre for Theoretical Studies in Nuclear Physics and Related Areas (ECT*) in Trento. The stimulating atmosphere of ECT* has provided the ideal environment to work on this project. I wish to express my sincere gratitude to Isabel Abril, Pablo de Vera, Jan Franz, Malgorzata Franz, Giovanni Garberoglio, Rafael Garcia-Molina, Gianluca Introzzi, and Simone Taioli for the numerous and illuminating discussions on the topics covered in the book and for their invaluable and stimulating comments. I would like to thank the students of my course entitled Computational Methods for Transport Phenomena (Department of Physics, University of Trento) for their clever questions and suggestions. I am also grateful to Maria Del Huerto Flammia for assisting me with the proofreading of this book.

Finally, warm thanks are due to my beloved children for their immense affection, my dear parents for their great love, and my cherished Roberta for her extraordinary patience.

Villazzano, January 2022   Maurizio Dapor

https://doi.org/10.1515/9783110675375-201

Contents

Preface —— VII

About the author —— XIII

Part I: Basic numerical and mathematical tools

1	**Basic numerical analysis** —— **3**	
1.1	Numerical differentiation —— 3	
1.2	Numerical quadrature —— 4	
1.2.1	Elementary quadrature formulas —— 4	
1.2.2	Gaussian quadrature —— 5	
1.2.3	The Monte Carlo method —— 6	
1.3	Ordinary differential equations —— 8	
1.3.1	Euler method —— 8	
1.3.2	Adams–Bashforth method —— 8	
1.3.3	Runge–Kutta method —— 9	
1.4	Linear second-order differential equations —— 10	
1.4.1	Numerov algorithm —— 10	

2	**Special functions of mathematical physics** —— **12**	
2.1	Legendre polynomials —— 12	
2.2	Associated Legendre functions —— 13	
2.3	Bessel functions —— 14	
2.4	Spherical harmonics —— 16	
2.5	Confluent hypergeometric functions —— 17	
2.6	Green function —— 18	

Part II: Quantum (non-relativistic) theory of elastic scattering and spin

3	**Partial wave expansion method** —— **21**	
3.1	Wave propagation, plane waves, and spherical waves —— 21	
3.2	Time-independent Schrödinger equation for the free particle —— 23	
3.3	Continuity equation —— 24	
3.4	Differential elastic scattering cross-section —— 25	
3.5	The radial equation —— 28	
3.6	Expansion of the plane wave —— 31	
3.7	Scattering amplitude —— 35	

3.8   Total elastic scattering cross-section and optical theorem —— 36
3.9   The first Born approximation —— 37
3.10   Rutherford elastic scattering cross-section —— 39

4   **Spectrum of angular momentum and spin** —— 41
4.1   Spectrum of angular momentum —— 41
4.2   Spin —— 46

5   **Phase shifts and atomic potential energy** —— 51
5.1   An important application of the Numerov algorithm: the eigenvalue
           problem —— 51
5.2   Phase-shift calculation —— 51
5.3   Atomic electron density and atomic potential energy —— 53
5.3.1   Thomas–Fermi model —— 53
5.3.2   Hartree and Hartree–Fock approximations —— 59

Part III: Quantum-relativistic equations and scattering

6   **The Klein–Gordon and the Dirac relativistic equations** —— 65
6.1   The natural system of units —— 65
6.2   The Lorentz transformation —— 66
6.3   Four-vectors and tensors —— 67
6.4   The Hamiltonian of a charged particle in an electromagnetic field —— 72
6.5   Klein–Gordon equation —— 73
6.6   Klein–Gordon particle in an electromagnetic field —— 74
6.7   Nonrelativistic limit of the Klein–Gordon equation —— 75
6.8   Difficulties of interpretation —— 77
6.9   Spin and gyromagnetic ratio of the electron —— 80
6.10   Dirac equation —— 81

7   **The Dirac equation and electron spin** —— 86
7.1   Manifestly covariant form of the Dirac equation —— 86
7.2   The Dirac equation and spin —— 88
7.2.1   Properties of commutators and anti-commutators —— 88
7.2.2   Spin —— 89
7.3   The solution of the Dirac equation for the free particle —— 91
7.4   The Pauli equation —— 94

8   **The Dirac theory of atoms** —— 97
8.1   Dirac equation in a central field —— 97
8.2   Dirac radial equations —— 102

8.3   The Dirac theory of one-electron atoms —— 104
8.4   The Dirac theory of one-electron atom energy levels —— 108
8.5   Many-electron atoms —— 109
8.5.1   Screening function —— 109
8.5.2   Corrections to the electrostatic potential —— 110

9   **Relativistic partial wave expansion method —— 112**
9.1   Scattering amplitudes —— 112
9.1.1   The fundamental equation —— 112
9.1.2   Effective Dirac potential —— 114
9.1.3   Phase shifts —— 115
9.1.4   Scattering amplitudes —— 115
9.2   Elastic scattering cross-section —— 118
9.2.1   Relativistic elastic scattering cross-section —— 118
9.2.2   Nonrelativistic limit —— 121
9.3   Phase-shift calculation —— 122
9.3.1   Lin, Sherman, and Percus transformation —— 122
9.3.2   Phase shifts —— 123
9.3.3   Numerical approach —— 125
9.4   Electron–molecule elastic scattering —— 128

10   **Density matrix and spin-polarization phenomena —— 130**
10.1   The density matrix —— 130
10.2   The spin-polarization —— 132
10.3   Polarization change following a collision —— 135
10.4   Polarization of an electron beam initially not polarized —— 138
10.5   Double elastic scattering —— 140
10.6   Change of the polarization in the general case —— 141
10.7   Sherman function for molecules —— 142

Part IV: Applications

11   **Exercises —— 147**
11.1   Exercise 1 —— 147
11.2   Exercise 2 —— 148
11.3   Exercise 3 —— 149
11.4   Exercise 4 —— 150
11.5   Exercise 5 —— 150
11.6   Exercise 6 —— 151
11.7   Exercise 7 —— 152
11.8   Exercise 8 —— 152

11.9 Exercise 9 —— 153
11.10 Exercise 10 —— 154
11.11 Exercise 11 —— 155
11.12 Exercise 12 —— 155
11.13 Exercise 13 —— 157
11.14 Exercise 14 —— 159
11.15 Exercise 15 —— 160
11.16 Exercise 16 —— 160
11.17 Exercise 17 —— 162
11.18 Exercise 18 —— 164
11.19 Exercise 19 —— 166
11.20 Exercise 20 —— 168
11.21 Exercise 21 —— 169
11.22 Exercise 22 —— 171
11.23 Exercise 23 —— 172
11.24 Exercise 24 —— 173
11.25 Exercise 25 —— 173
11.26 Exercise 26 —— 174
11.27 Exercise 27 —— 175

Bibliography —— 177

Index —— 179

About the author

Maurizio Dapor

Senior researcher at the European Centre for Theoretical Studies in Nuclear Physics and Related Areas (FBK, Trento, Italy). Ph.D.: Materials Science and Engineering; M.Sc.: Physics. Italian National Scientific Habilitation to Full Professor in Theoretical Physics of Matter; Teaching Fellow: Department of Physics, University of Trento, Italy; Visiting Professor: Gdańsk University of Technology, Poland (2021), Department of Applied Physics, University of Alicante, Spain (2016), Department of Materials Science and Engineering, University of Sheffield, UK (2015); Scientific Consultant: Integrated Systems Laboratory, Swiss Federal Institute of Technology (ETH), Zurich (2009). Member of the American Physical Society, the European Physical Society, and the Italian Physical Society. Author of the books: *Electron–Beam Interactions with Solids*, Springer Tracts in Modern Physics 186, Springer 2003, *Teoria della Relatività*, Zanichelli 2008, *Relatività e Meccanica Quantistica Relativistica*, Carocci 2011, and *Transport of Energetic Electrons in Solids*, Springer Tracts in Modern Physics 271, 3rd Edition, Springer 2020.

https://doi.org/10.1515/9783110675375-202

Part I: **Basic numerical and mathematical tools**

1 Basic numerical analysis

This chapter is devoted to the basic concepts of numerical analysis used in computational physics [4, 6, 15, 20]. It deals with the numerical tools that are central for the computational modeling of physical systems. We will introduce the main methods of numerical differentiation and numerical quadrature (or integration) (trapezoidal rule, Simpson rule, Bode rule, Gaussian quadrature, and the Monte Carlo method). Furthermore, we will describe important methods for the solution of ordinary differential equations (the Euler method, Adams–Bashforth method, and Runge–Kutta method). We will describe, at the end, the Numerov algorithm for solving linear second-order differential equations.

1.1 Numerical differentiation

To discuss numerical differentiation, we must remind our readers that any function $f(x)$, infinitely differentiable in 0, can be expressed as a power series (Maclaurin series):

$$f(x) = \sum_{n=0}^{\infty} \frac{x^n}{n!} f^{(n)}(0) = f(0) + xf'(0) + \frac{x^2}{2!} f''(0) + \cdots, \tag{1.1}$$

where $f^{(n)}(0)$ represents the nth derivative of $f(x)$ calculated in the origin. For example, $f^{(1)}(0) = f'(0)$. Since zero is the point where we have calculated the derivatives, the Maclaurin series is a special case of a more general expression. In fact, if the function $f(x)$ is infinitely differentiable at a point a, then it can be expressed as the following power series (Taylor series):

$$f(x) = \sum_{n=0}^{\infty} \frac{(x-a)^n}{n!} f^{(n)}(a) = f(a) + (x-a)f'(a) + \frac{(x-a)^2}{2!} f''(a) + \cdots, \tag{1.2}$$

so that the Maclaurin series is the Taylor series for $a = 0$. If $x \ll 1$, the Maclaurin expansion can be truncated so that $f(x)$ can be approximated with a polynomial, for example, a second-order one:

$$f(x) \approx f(0) + xf'(0) + \frac{x^2}{2!} f''(0). \tag{1.3}$$

The higher the order of the polynomial, of course, the better the approximation. Anyway, all the neglected terms in the previous equation are smaller and smaller since they are proportional to powers of x, a number much smaller than 1. In the neighborhood of $x = 0$, we thus have, for any $h \ll 1$,

$$f(h) = f(0) + hf'(0) + \frac{h^2}{2!} f''(0) + \cdots, \tag{1.4}$$

https://doi.org/10.1515/9783110675375-001

and

$$f(-h) = f(0) - hf'(0) + \frac{h^2}{2!}f''(0) + \cdots, \tag{1.5}$$

so that

$$f' = \frac{f(h) - f(-h)}{2h} + \mathcal{O}(h^2). \tag{1.6}$$

Note that, in Eq. (1.6), we have indicated with $\mathcal{O}(h^2)$ all the terms of order equal or greater than 2. The formula expressed by Eq. (1.6) is known as the "three-point" formula. The tree-point formula is exact if f is a second-degree polynomial function in the interval $[-h, h]$. In other words, Eq. (1.6) assumes a quadratic polynomial interpolation of the function f through $x = -h$, $x = 0$, and $x = h$. A "two point" formula can be obtained using only Eq. (1.4):

$$f' = \frac{f(h) - f(0)}{h} + \mathcal{O}(h), \tag{1.7}$$

$\mathcal{O}(h)$ includes all the terms of order equal or greater than 1. The "two-point" formula is exact if f is a linear function in the interval $[0, h]$. In other words Eq. (1.7) assumes a linear interpolation of the function f through $x = 0$ and $x = h$. Using both Eqs. (1.4) and (1.5), we can easily obtain the second-order derivative:

$$f'' = \frac{f(h) - 2f(0) + f(-h)}{h^2} + \mathcal{O}(h^2). \tag{1.8}$$

Higher-order derivatives can of course be obtained using similar procedures.

1.2 Numerical quadrature

1.2.1 Elementary quadrature formulas

Let us consider a small interval $[0, h]$ where we can assume that the function f is approximately linear. The so-called trapezoidal rule immediately follows:

$$\int_0^h f(x)\, dx = \frac{f(0) + f(h)}{2} h + \mathcal{O}(h^3). \tag{1.9}$$

The name "trapezoidal rule" originates from the fact that, if f has positive values, the integral is approximated by the area of a trapezoid. The same rule can also be expressed in a symmetric form about $x = 0$. To do that, let us assume that f is approximately linear in both the two intervals $[-h, 0]$ and $[0, h]$), so that we can write

$$\int_{-h}^{h} f(x)\, dx = \frac{f(-h) + 2f(0) + f(h)}{2} h + \mathcal{O}(h^3). \tag{1.10}$$

To increase the accuracy, let us note that, for $|x| < h$, we can write

$$f(x) \approx f(0) + \frac{f(h) - f(-h)}{2h} x + \frac{f(h) - 2f(0) + f(-h)}{2h^2} x^2. \tag{1.11}$$

We can now integrate this equation from $-h$ to h to obtain

$$\int_{-h}^{h} f(x)\, dx = 2hf(0) + \frac{2h^3}{3} \frac{f(h) - 2f(0) + f(-h)}{2h^2} + \mathcal{O}(h^5). \tag{1.12}$$

The so-called Simpson quadrature rule can be deduced from the last equation:

$$\int_{-h}^{h} f(x)\, dx = \frac{f(h) + 4f(0) + f(-h)}{3} h + \mathcal{O}(h^5). \tag{1.13}$$

Higher-order quadrature rules can be obtained including in the calculation further terms of the Taylor series. In particular, we can easily obtain the so-called Bode quadrature rule:

$$\int_{x_0}^{x_0+4h} f(x)\, dx$$
$$= \frac{2h}{45} [7f(x_0) + 32f(x_0 + h) + 12f(x_0 + 2h)$$
$$+ 32f(x_0 + 3h) + 7f(x_0 + 4h)] + \mathcal{O}(h^7). \tag{1.14}$$

1.2.2 Gaussian quadrature

The Gaussian quadrature rule is another way to perform integration of functions. It is a very fast algorithm, but it requires that the integrand be a smooth function of x. To describe this method, let us consider the integral:

$$S = \int_{-1}^{1} f(x) dx. \tag{1.15}$$

All the elementary quadrature formulas just described use the following approach to numerically approximate S:

$$S \approx \sum_{n=1}^{N} c_n f(x_n), \qquad (1.16)$$

where

$$x_n = 2\frac{n-1}{N-1} - 1, \qquad (1.17)$$

and the coefficients c_n are obtained by solving the following system of N linear equations:

$$\int_1^1 x^p \, dx = \sum_{n=1}^{N} c_n x_n^p, \qquad (1.18)$$

with $p = 0,1,\ldots,N-1$. In the case of Simpson's rule, for example, $N = 3$, $x_1 = -1$, $x_2 = 0$, $x_3 = +1$, and, once solved the system of linear equations (1.18), $c_1 = 1/3$, $c_2 = 4/3$, $c_3 = 1/3$. If we give up the requirement of equally spaced points x_n, choosing x_n as the N zeros of the Legendre polynomial $P_N(x)$ (see next chapter), we have

$$c_n = \frac{2}{(1 - x_n^2)[P_N'(x_n)]^2}. \qquad (1.19)$$

For example, if $N = 3$, Gaussian quadrature corresponds to the use of Eq. (1.16) with $x_1 = \sqrt{3/5}$, $x_2 = 0$, $x_3 = -\sqrt{3/5}$, $c_1 = 5/9$, $c_2 = 8/9$, and $c_3 = 5/9$. The abscissae x_n and the weights c_n for higher values of N can be found in the literature or directly calculated. Also note that, to compute the integral:

$$S_{ab} = \int_a^b f(x) \, dx, \qquad (1.20)$$

the following variable change has to be carried out:

$$y = 2\frac{x-a}{b-a} - 1. \qquad (1.21)$$

Gaussian quadrature can still be used for rapidly varying integrand functions by integrating over many subintervals of the integration range.

1.2.3 The Monte Carlo method

The Monte Carlo method is a very efficient way, based on the generation of random numbers, to evaluate multi-dimensional integrals, but, to illustrate the use of random numbers to calculate integrals, let us just consider the one-dimensional case. Assuming that N abscissae x_i are chosen at random in the closed interval $[0, 1]$, we can write

$$\int_0^1 f(x)\,dx \approx \frac{1}{N} \sum_{i=1}^{N} f(x_i). \tag{1.22}$$

This equation represents the Monte Carlo method, expressed in its simplest form, for integrating a function f from 0 to 1. Please note that the values of the N random abscissae have to be chosen with equal probability in the range $[0, 1]$ or, in other words, they must be random numbers uniformly distributed within the interval $[0, 1]$. The efficiency of the Monte Carlo quadrature can be improved by introducing a new positive function $w(x)$ satisfying the condition:

$$\int_0^1 w(x)\,dx = 1. \tag{1.23}$$

Let us now consider the function $y(x)$ defined as

$$y(x) = \int_0^x w(x')\,dx' \tag{1.24}$$

and observe that

$$\frac{dy}{dx} = w(x), \tag{1.25}$$
$$y(x = 0) = 0, \tag{1.26}$$
$$y(x = 1) = 1. \tag{1.27}$$

Since

$$\int_0^1 f(x)\,dx = \int_0^1 \frac{f(x)}{w(x)} w(x)\,dx = \int_0^1 \frac{f(x)}{w(x)} \frac{dy}{dx}\,dx = \int_0^1 \frac{f(x(y))}{w(x(y))}\,dy, \tag{1.28}$$

we have

$$\int_0^1 f(x)\,dx \approx \frac{1}{N} \sum_{i=1}^{N} \frac{f(x(y_i))}{w(x(y_i))}. \tag{1.29}$$

Using the latter equation, the idea is to choose a function $w(x)$ that behaves approximately like $f(x)$. In such a way, little computing power is used where both $w(x)$ and $f(x)$ are small and, as a consequence, x values are less important.

1.3 Ordinary differential equations

Let us consider the simplest ordinary differential equation:

$$\frac{dy}{dx} = f(x,y).\tag{1.30}$$

In this equation, $f(x,y)$ is a known function of x and y. We wish to find the function $y(x)$ satisfying Eq. (1.30) and assuming the following initial condition:

$$y(x_0) = y_0,\tag{1.31}$$

where x_0 and y_0 are real numbers.

1.3.1 Euler method

Let us now describe the Euler method. We wish to find the value of the function y for a given value of the variable x, for example, $x = 1$. Let us consider the case where x ranges in the interval $[0,1]$ so that $x_0 = 0$ and $y_0 = y(0)$. Let us divide the interval $[0,1]$ in N subintervals of length $h = 1/N$, and let us indicate with y_n the approximated value of $y(x_n)$, where $x_n = nh$:

$$y_n = y(x_n).\tag{1.32}$$

Thus, the following simple recursion formula numerically solves Eq. (1.30) by a step-by-step integration:

$$y_{n+1} = y_n + hf(x_n, y_n) + \mathcal{O}(h^2),\tag{1.33}$$

Note that the local error is $\mathcal{O}(h^2)$. To calculate the global error, we have to consider the N steps necessary to integrate from $x = 0$ to $x = 1$. As a consequence, the global error is given by $N\,\mathcal{O}(h^2) = (1/h)\,\mathcal{O}(h^2) \approx \mathcal{O}(h)$.

1.3.2 Adams–Bashforth method

This method is based on the exact integration:

$$y_{n+1} = y_n + \int_{x_n}^{x_{n+1}} f(x,y)\,dx.\tag{1.34}$$

By linear extrapolation we obtain the approximate formula:

$$f \approx f(x_{n-1}, y_{n-1}) + \frac{f(x_n, y_n) - f(x_{n-1}, y_{n-1})}{x_n - x_{n-1}}(x - x_{n-1}) \tag{1.35}$$

or, equivalently,

$$f = \frac{x - x_{n-1}}{h} f_n - \frac{x - x_n}{h} f_{n-1} + \mathcal{O}(h^2), \tag{1.36}$$

where $f_n = f(x_n, y_n)$. From

$$\int_{x_n}^{x_{n+1}} \frac{x - x_{n-1}}{h} f_n \, dx = \frac{f_n}{2}(x_{n+1} + x_n - 2x_{n-1}) = \frac{3f_n}{2} h, \tag{1.37}$$

$$\int_{x_n}^{x_{n+1}} \frac{x - x_n}{h} f_{n-1} \, dx = \frac{f_{n-1}}{2}(x_{n+1} + x_n - 2x_n) = \frac{f_{n-1}}{2} h, \tag{1.38}$$

we obtain

$$y_{n+1} = y_n + \frac{h}{2}(3f_n - f_{n-1}) + \mathcal{O}(h^3). \tag{1.39}$$

Equation (1.39) represents the Adams–Bashforth two-step method.

1.3.3 Runge–Kutta method

A more accurate approach is represented by the so-called Runge–Kutta method. According to the fourth-order Runge–Kutta method, we have

$$y_{n+1} = y_n + \frac{1}{6}(k_1 + 2k_2 + 2k_3 + k_4) + \mathcal{O}(h^5), \tag{1.40}$$

where

$$\begin{aligned}
k_1 &= hf(x_n, y_n), \\
k_2 &= hf(x_n + h/2, y_n + k_1/2), \\
k_3 &= hf(x_n + h/2, y_n + k_2/2), \\
k_4 &= hf(x_n + h, y_n + k_3).
\end{aligned} \tag{1.41}$$

1.4 Linear second-order differential equations

A linear differential equation of the second-order has the form:

$$\frac{d^2y}{dx^2} + f^2(x)y = g(x).$$ (1.42)

Let us consider $y(x)$. If $h \ll 1$ we can expand it in a Maclaurin series as follows:

$$y(h) = y_0 + hy' + \frac{h^2}{2!}y'' + \frac{h^3}{3!}y''' + \frac{h^4}{4!}y'''' + \frac{h^5}{5!}y''''' + \mathcal{O}(h^6),$$ (1.43)

where y_0 is the value assumed by the function at $x = 0$:

$$y_0 = y(0).$$ (1.44)

Please note that all the derivatives are calculated at $x = 0$. We can also write

$$y(-h) = y_0 - hy' + \frac{h^2}{2!}y'' - \frac{h^3}{3!}y''' + \frac{h^4}{4!}y'''' - \frac{h^5}{5!}y''''' + \mathcal{O}(h^6),$$ (1.45)

and, as a consequence,

$$\frac{y(h) - 2y(0) + y(-h)}{h^2} = y'' + \frac{h^2}{12}y'''' + \mathcal{O}(h^4).$$ (1.46)

Hence, we can also write

$$\frac{y_{n+1} - 2y_n + y_{n-1}}{h^2} = y_n'' + \frac{h^2}{12}y_n'''' + \mathcal{O}(h^4),$$ (1.47)

where

$$y_n = y(x_n),$$ (1.48)
$$x_n = nh,$$ (1.49)

and all the derivatives are calculated at $x = x_n$.

1.4.1 Numerov algorithm

Many differential equations in physics have the simple form of Eq. (1.42), i. e., they are linear differential equations of the second order. From Eq. (1.42) it follows that

$$y_n'''' = \frac{d^2}{dx^2}(-f^2y + g)|_{x=x_n}.$$ (1.50)

Keeping in mind Eq. (1.47), we thus obtain, on the one hand,

$$y_n'''' = -\frac{f_{n+1}^2 y_{n+1} - 2f_n^2 y_n + f_{n-1}^2 y_{n-1}}{h^2} + \frac{g_{n+1} - 2g_n + g_{n-1}}{h^2} + \mathcal{O}(h^2). \qquad (1.51)$$

On the other hand, since

$$y_n'' = -f_n^2 y_n + g_n, \qquad (1.52)$$

we have

$$\frac{y_{n+1} - 2y_n + y_{n-1}}{h^2} = -f_n^2 y_n + g_n + \frac{h^2}{12} y_n'''' + \mathcal{O}(h^4). \qquad (1.53)$$

Using Eqs. (1.51) and (1.53), we obtain

$$\left(\frac{12}{h^2} + f_{n+1}^2\right) y_{n+1} + \left(-\frac{24}{h^2} + 10 f_n^2\right) y_n + \left(\frac{12}{h^2} + f_{n-1}^2\right) y_{n-1}$$
$$= g_{n+1} + 10 g_n + g_{n-1} + \mathcal{O}(h^4),$$

and, after a few simple algebraical manipulations, the Numerov algorithm:

$$\left(1 + \frac{h^2}{12} f_{n+1}^2\right) y_{n+1} - 2\left(1 - \frac{5h^2}{12} f_n^2\right) y_n + \left(1 + \frac{h^2}{12} f_{n-1}^2\right) y_{n-1}$$
$$= \frac{h^2}{12}(g_{n+1} + 10 g_n + g_{n-1}) + \mathcal{O}(h^6). \qquad (1.54)$$

2 Special functions of mathematical physics

This chapter is an introduction to the functions of mathematical physics useful for our aims [1, 11, 15]. This mathematical introduction will help our readers to understand the concepts that we will present later in the book. We will introduce, in particular, recursion relations useful for determining numerically the main functions of mathematical physics (i. e., Legendre polynomials, associated Legendre functions, and regular and irregular spherical Bessel functions). We will discuss spherical harmonics and introduce the confluent hypergeometric function and the Green function.

2.1 Legendre polynomials

We are interested in the Legendre polynomials because they are (as we will show in the next chapter) eigenfunctions of the square of the orbital angular momentum:

$$\mathbf{L}^2 = -\hbar^2 \left(\frac{\partial^2}{\partial \vartheta^2} + \cot \vartheta \frac{\partial}{\partial \vartheta} + \frac{1}{\sin^2 \vartheta} \frac{\partial^2}{\partial \varphi^2} \right), \tag{2.1}$$

$$\mathbf{L}^2 P_l(\cos \vartheta) = \hbar^2 l(l+1) P_l(\cos \vartheta). \tag{2.2}$$

They are the following polynomials of degree l of the variable u ($-1 \le u \le 1$):

$$P_l(u) = \frac{1}{2^l l!} \frac{d^l}{du^l} (u^2 - 1)^l, \tag{2.3}$$

where $l = 0, 1, 2, \ldots$, and satisfy the equation:

$$\left[(1 - u^2) \frac{d^2}{du^2} - 2u \frac{d}{du} + l(l+1) \right] P_l(u) = 0. \tag{2.4}$$

The following recursion relationships, valid also for $l = 0$, once defined $P_{-1} = 1$, can be used to calculate Legendre polynomials:

$$(l+1) P_{l+1}(u) + l P_{l-1}(u) = (2l+1) u P_l(u), \tag{2.5}$$

$$(1 - u^2) \frac{d}{du} P_l(u) = l P_{l-1}(u) - l u P_l(u). \tag{2.6}$$

The first five Legendre polynomials can be directly calculated as follows:

$$P_0 = 1, \tag{2.7}$$

$$P_1 = u, \tag{2.8}$$

$$P_2 = \frac{1}{2}(3u^2 - 1), \tag{2.9}$$

https://doi.org/10.1515/9783110675375-002

$$P_3 = \frac{1}{2}(5u^3 - 3u),\tag{2.10}$$

$$P_4 = \frac{1}{8}(35u^4 - 30u^2 + 3).\tag{2.11}$$

Please note that

$$P_l(1) = 1,\tag{2.12}$$

$$P_l(-1) = (-1)^l.\tag{2.13}$$

The Legendre polynomials satisfy the closure relationship:

$$\frac{1}{2}\sum_{l=0}^{\infty}(2l+1)P_l(u)P_l(v) = \delta(u-v),\tag{2.14}$$

where δ is the Dirac delta distribution.

2.2 Associated Legendre functions

The associated Legendre functions are given by

$$P_l^m(u) = (1-u^2)^{\frac{1}{2}m}\frac{d^m}{du^m}P_l(u),\tag{2.15}$$

where $m = 0,1,2,\ldots,l$. They satisfy the equation:

$$\left[(1-u^2)\frac{d^2}{du^2} - 2u\frac{d}{du} + l(l+1) - \frac{m^2}{1-u^2}\right]P_l^m(u) = 0.\tag{2.16}$$

The Legendre polynomials are the particular associated Legendre functions corresponding to $m = 0$:

$$P_l(u) = P_l^0(u).\tag{2.17}$$

The associated Legendre functions satisfy the orthonormality relationship:

$$\int_{-1}^{+1}P_k^m(u)P_l^m(u)\,du = \frac{2}{2l+1}\frac{(l+m)!}{(l-m)!}\delta_{kl},\tag{2.18}$$

where δ_{kl} is the Kronecker symbol (equal to 1 if $k = l$ and to 0 otherwise) and, as a consequence, the Legendre polynomials satisfy the orthonormality relationship:

$$\int_{-1}^{+1}P_k(u)P_l(u)\,du = \frac{2}{2l+1}\delta_{kl}.\tag{2.19}$$

To numerically calculate the associated Legendre functions, the following recursion relationships can be used:

$$(l - m + 1)P_{l+1}^m(u) + (l + m)P_{l-1}^m(u) = (2l + 1)uP_l^m(u), \tag{2.20}$$

$$(1 - u^2)\frac{d}{du}P_l^m(u) = (l + m)P_{l-1}^m(u) - luP_l^m(u). \tag{2.21}$$

2.3 Bessel functions

The Bessel equation of order v satisfies the equation:

$$x^2\frac{d^2y}{dx^2} + x\frac{dy}{dx} + (x^2 - v^2)y = 0. \tag{2.22}$$

Any linear combination of the Bessel functions J_{-v} and J_{+v} is a solution to this equation. The Schrödinger equation of a particle in a constant potential V_0 can be written as

$$(\nabla^2 + K^2 - U_0)\Psi = 0, \tag{2.23}$$

where $\mathbf{K} = \mathbf{p}/m$, $K^2 = 2mE/\hbar^2$, and $U_0 = 2mV_0/\hbar^2$. Let us now expand the wave function in Legendre polynomials:

$$\Psi(r, \cos\vartheta) = \sum_{l=0}^{\infty} a_l\frac{y_l(r)}{r}P_l(\cos\vartheta). \tag{2.24}$$

Since, in spherical coordinates,

$$\nabla^2 = \frac{\partial^2}{\partial r^2} + \frac{2}{r}\frac{\partial}{\partial r} + \frac{1}{r^2}\left(\frac{\partial^2}{\partial\vartheta^2} + \cot\vartheta\frac{\partial}{\partial\vartheta} + \frac{1}{\sin^2\vartheta}\frac{\partial^2}{\partial\varphi^2}\right) = \frac{\partial^2}{\partial r^2} + \frac{2}{r}\frac{\partial}{\partial r} - \frac{1}{r^2}\frac{\mathbf{L}^2}{\hbar^2}, \tag{2.25}$$

taking into account Eq. (2.2), the Schrödinger equation becomes

$$\sum_{l=0}^{\infty}\left[\frac{\partial^2}{\partial r^2} + \frac{2}{r}\frac{\partial}{\partial r} - \frac{l(l+1)}{r^2} + K^2 - U_0\right]a_l\frac{y_l(r)}{r}P_l(\cos\vartheta) = 0. \tag{2.26}$$

All the coefficients of this expansion must then satisfy the following differential equation:

$$\left[\frac{\partial^2}{\partial r^2} + \frac{2}{r}\frac{\partial}{\partial r} - \frac{l(l+1)}{r^2} + K^2 - U_0\right]a_l\frac{y_l(r)}{r} = 0. \tag{2.27}$$

Let us now introduce

$$k^2 = K^2 - U_0, \tag{2.28}$$

so that

$$\left[\frac{d^2}{dr^2} - \frac{l(l+1)}{r^2} + k^2\right]y_l(r) = 0. \tag{2.29}$$

If $x = kr$, the last equation can be rewritten as

$$\left[\frac{d^2}{dx^2} - \frac{l(l+1)}{x^2} + 1\right]y_l(x) = 0. \tag{2.30}$$

Let us now introduce the spherical Bessel function of order l. It is defined by

$$j_l(x) = \sqrt{\frac{\pi}{2x}}J_{l+1/2}(x). \tag{2.31}$$

Note that $x^{1/2}J_{l+1/2}$ is a solution to Eq. (2.30). Indeed

$$\left[\frac{d^2}{dx^2} - \frac{l(l+1)}{x^2} + 1\right]x^{1/2}J_{l+1/2}$$

$$= \sqrt{x}\left\{\frac{d^2J_{l+1/2}}{dx^2} + \frac{1}{x}\frac{dJ_{l+1/2}}{dx} + \left[1 - \frac{(l+1/2)^2}{x^2}\right]J_{l+1/2}\right\}$$

$$= x^{-3/2}\left\{x^2\frac{d^2J_{l+1/2}}{dx^2} + x\frac{dJ_{l+1/2}}{dx} + [x^2 - (l+1/2)^2]J_{l+1/2}\right\}, \tag{2.32}$$

and $J_{l+1/2}$ is a solution of the Bessel equation (2.22) with $v = l+1/2$. As a consequence, the function $krj_l(kr)$ is a solution to Eq. (2.29). Defining now the spherical Neumann function of order l (or irregular spherical Bessel function of order l) as

$$n_l(x) = (-1)^{l+1}\sqrt{\frac{\pi}{2x}}J_{-l-1/2}(x), \tag{2.33}$$

we see that $krn_l(kr)$ is also a solution of Eq. (2.29). The first three regular and irregular spherical Bessel functions are given by

$$j_0 = \frac{\sin x}{x}, \tag{2.34}$$

$$j_1 = \frac{\sin x}{x^2} - \frac{\cos x}{x}, \tag{2.35}$$

$$j_2 = \left(\frac{3}{x^3} - \frac{1}{x}\right)\sin x - \frac{3}{x^2}\cos x, \tag{2.36}$$

$$n_0 = -\frac{\cos x}{x}, \tag{2.37}$$

$$n_1 = -\frac{\cos x}{x^2} - \frac{\sin x}{x}, \tag{2.38}$$

$$n_2 = \left(-\frac{3}{x^3} + \frac{1}{x}\right)\cos x - \frac{3}{x^2}\sin x. \tag{2.39}$$

The following equations hold:

$$\overset{\flat}{j_l}(x) \underset{x\to 0}{\sim} \frac{x^l}{1\cdot 3\cdot \,\cdots\, \cdot(2l+1)}\,, \tag{2.40}$$

$$n_l(x) \underset{x\to 0}{\sim} -\frac{1\cdot 3\cdot \,\cdots\, \cdot(2l-1)}{x^{l+1}}\,, \tag{2.41}$$

$$j_l(0) = \delta_{l0}\,. \tag{2.42}$$

Furthermore, the Bessel and Neumann functions have the following asymptotic behavior:

$$j_l(x) \underset{x\to\infty}{\sim} \frac{\sin(x - l\pi/2)}{x}\,, \tag{2.43}$$

$$n_l(x) \underset{x\to\infty}{\sim} -\frac{\cos(x - l\pi/2)}{x}\,. \tag{2.44}$$

To numerically calculate the Bessel and Neumann functions, the following recursion relationship can be used:

$$xf_{l-1} - (2l+1)f_l + xf_{l+1} = 0\,, \tag{2.45}$$

$$xf_{l-1} - (l+1)f_l - x\frac{df_l}{dx} = 0\,, \tag{2.46}$$

where f_l is any linear combination of the functions j_l and n_l.

2.4 Spherical harmonics

The spherical harmonics Y_l^m are the eigenfunctions common to the operators $\mathbf{L}^2$ and L_z. With $l = 0, 1, \ldots$ and $m = -l, -l+1, \ldots, l-1, l$, we can write

$$\mathbf{L}^2 Y_l^m = \hbar^2 l(l+1) Y_l^m \tag{2.47}$$

and

$$L_z Y_l^m = \hbar m Y_l^m\,.^{[1]} \tag{2.48}$$

If $m \geq 0$, the spherical harmonics are given by the following equation:

$$Y_l^m(\vartheta, \varphi) = (-1)^m \sqrt{\frac{(2l+1)(l-m)!}{4\pi(l+m)!}} P_l^m(\cos\vartheta) \exp(im\varphi)\,. \tag{2.49}$$

---

[1] Please note that here we use the symbol m to indicate the eigenvalues of L_z divided by $\hbar$. Consideration of the context always allows one to avoid confusing it with the electron mass, also indicated with m elsewhere.

If $m < 0$, they can be obtained by

$$(-1)^m Y_l^{-m}(\vartheta, \varphi) = Y_l^{m*}(\vartheta, \varphi). \tag{2.50}$$

Please note that the spherical harmonics are normalized to unity on the unit sphere and satisfy the orthonormality and closure relationship. They form a complete set of square-integrable orthonormal functions on the unit sphere:

$$\int_0^{2\pi} d\varphi \int_0^\pi \sin\vartheta \, d\vartheta \, Y_j^{k*}(\vartheta, \varphi) Y_l^m(\vartheta, \varphi) = \delta_{km} \delta_{jl}, \tag{2.51}$$

$$\sum_{l=0}^\infty \sum_{m=-l}^l Y_l^{m*}(\vartheta, \varphi) Y_l^m(\vartheta', \varphi') = \delta(\Omega - \Omega') = \frac{\delta(\vartheta - \vartheta')\delta(\varphi - \varphi')}{\sin\vartheta}. \tag{2.52}$$

2.5 Confluent hypergeometric functions

Another important equation is the Kummer equation (also known as the Laplace equation):

$$x\frac{d^2y}{dx^2} + (b - x)\frac{dy}{dx} - ay = 0. \tag{2.53}$$

The regular solution of this equation is the confluent hypergeometric function:

$$y = M(a, b, x) = 1 + \frac{(a)_1}{(b)_1} x + \frac{(a)_2}{(b)_2} \frac{x^2}{2!} + \cdots + \frac{(a)_n}{(b)_n} \frac{x^n}{n!} + \cdots, \tag{2.54}$$

where

$$(a)_n = a(a + 1)(a + 2) \cdots (a + n - 1), \tag{2.55}$$
$$(b)_n = b(b + 1)(b + 2) \cdots (b + n - 1), \tag{2.56}$$

and

$$(a)_0 = 1, \tag{2.57}$$
$$(b)_0 = 1, \tag{2.58}$$

so that

$$\begin{aligned}
M(a, b, x) = 1 &+ \frac{a}{b} x + \frac{a(a + 1)}{b(b + 1)} \frac{x^2}{2!} \\
&+ \frac{a(a + 1)(a + 2)}{b(b + 1)(b + 2)} \frac{x^3}{3!} + \cdots \\
&+ \frac{a(a + 1) \cdots (a + n - 1)}{b(b + 1) \cdots (b + n - 1)} \frac{x^n}{n!} + \cdots.
\end{aligned} \tag{2.59}$$

The Γ function has the following property:

$$\Gamma(a+1) = a\Gamma(a).\tag{2.60}$$

Thus we have

$$\Gamma(a+2) = (a+1)\Gamma(a+1) = (a+1)a\Gamma(a),\tag{2.61}$$
$$\Gamma(a+3) = (a+2)\Gamma(a+2) = (a+2)(a+1)a\Gamma(a),\tag{2.62}$$

and so on. As a consequence, the confluent hypergeometric function can also be expressed as

$$M(a,b,x) = \sum_{n=0}^{\infty} \frac{[\Gamma(n+a)/\Gamma(a)]}{[\Gamma(n+b)/\Gamma(b)]} \frac{x^n}{n!}.\tag{2.63}$$

2.6 Green function

The operator $\nabla^2 + k^2$ satisfies the equation:

$$(\nabla^2 + k^2)g(\mathbf{r},\mathbf{r}') = \delta(\mathbf{r}-\mathbf{r}'),\tag{2.64}$$

where

$$\delta(\mathbf{r}-\mathbf{r}')$$

is the Dirac delta function and

$$g(\mathbf{r},\mathbf{r}') = -\frac{\exp(ik|\mathbf{r}-\mathbf{r}'|)}{4\pi|\mathbf{r}-\mathbf{r}'|}\tag{2.65}$$

is the Green function.

Part II: **Quantum (non-relativistic) theory of elastic
scattering and spin**

3 Partial wave expansion method

The theory of the electron–atom scattering processes can be presented in a very simple manner assuming the following approximation. Since the masses of the target atoms are much larger than those of the incident electrons, the target atoms can be considered, as a first approximation, as infinitely massive and at rest. Neglecting their structure, the problem can be further simplified and reduced to the calculation of the differential and total elastic scattering cross-sections of a beam of electrons (with a given velocity) by a fixed center of force (described by a central potential). This is the approach we present in this chapter [2, 5, 11, 14, 18, 22, 26, 31, 32].

3.1 Wave propagation, plane waves, and spherical waves

The function $F(x, t)$, where x is the position and t is the time, represents a wave that propagates along the x-axis with velocity v if

$$F(x, t) = F(x - vt).$$
(3.1)

Our readers can easily verify that it satisfies the d'Alembert equation, i. e.,

$$\frac{\partial^2 F}{\partial x^2} - \frac{1}{v^2} \frac{\partial^2 F}{\partial t^2} = 0.$$
(3.2)

By definition, a plane wave is a function of x and t that assumes, for any given x and t, the same value all along the plane perpendicular to the x-axis passing through x. So, $F(x, t) = F(x - vt)$ is a plane wave. A periodical plane wave can be expressed as:

$$F(x, t) = f \exp\left[i \frac{2\pi}{\lambda}(x - vt)\right] = f \exp[i(kx - \omega t)],$$
(3.3)

where f is the amplitude, λ the wavelength, $k = 2\pi/\lambda$ the wavenumber, and $\omega = 2\pi v/\lambda$ the angular frequency. By definition, a spherical wave is a function G of the modulus r of the radius vector $\mathbf{r}$ and of time t:

$$G = G(r, t)$$
(3.4)

that satisfies the three-dimensional version of the d'Alembert equation:

$$\nabla^2 G - \frac{1}{v^2} \frac{\partial G^2}{\partial t^2} = 0.$$
(3.5)

Here, ∇^2 is the Laplacian operator, which, in spherical coordinates, is given by

$$\nabla^2 = \frac{\partial^2}{\partial r^2} + \frac{2}{r} \frac{\partial}{\partial r} + \frac{1}{r^2} \Lambda,$$
(3.6)

https://doi.org/10.1515/9783110675375-003

where operator Λ depends only on the angular variables ϑ and φ:

$$\Lambda = \frac{\partial^2}{\partial\vartheta^2} + \frac{\cos\vartheta}{\sin\vartheta}\frac{\partial}{\partial\vartheta} + \frac{1}{\sin^2\vartheta}\frac{\partial^2}{\partial\varphi^2}. \tag{3.7}$$

Since the function G depends on the modulus r of $\mathbf{r}$, it does not depend on the angular variables ϑ and φ. As a consequence the d'Alembert equation reads

$$\frac{\partial^2 G}{\partial r^2} + \frac{2}{r}\frac{\partial G}{\partial r} - \frac{1}{v^2}\frac{\partial^2 G}{\partial t^2} = 0. \tag{3.8}$$

Please note that

$$\frac{\partial}{\partial r}(rG) = G + r\frac{\partial G}{\partial r}, \tag{3.9}$$

and

$$\frac{\partial^2}{\partial r^2}(rG) = \frac{\partial}{\partial r}\left(G + r\frac{\partial G}{\partial r}\right) = 2\frac{\partial G}{\partial r} + r\frac{\partial^2 G}{\partial r^2}. \tag{3.10}$$

As a consequence,

$$\frac{\partial^2}{\partial r^2}(rG) - \frac{1}{v^2}\frac{\partial^2}{\partial t^2}(rG) = 0. \tag{3.11}$$

In other words, the function rG satisfies the d'Alembert equation in one dimension, Eq. (3.2), so that

$$rG(r,t) = F(r - vt) \tag{3.12}$$

or

$$G(r,t) = \frac{F(r - vt)}{r}, \tag{3.13}$$

where $F(r - vt)$ is a plane wave propagating with velocity v. If $F(r - vt)$ is a periodical plane wave, then

$$f(\vartheta,\varphi)\,G(r,t) = f(\vartheta,\varphi)\frac{\exp[i(kr - \omega t)]}{r} = \exp(-i\omega t)\,\psi_s(r,\vartheta,\varphi), \tag{3.14}$$

where

$$\psi_s(r,\vartheta,\varphi) = f(\vartheta,\varphi)\frac{\exp(ikr)}{r}. \tag{3.15}$$

3.2 Time-independent Schrödinger equation for the free particle

We remind our readers that, in quantum mechanics, energy E and momentum $\mathbf{p}$ are represented by differential operators according to the following correspondence rules:

$$E \rightarrow i\hbar \frac{\partial}{\partial t}, \tag{3.16}$$

$$\mathbf{p} \rightarrow \frac{\hbar}{i} \nabla, \tag{3.17}$$

where $\hbar = h/2\pi$, h is Planck's constant and, in Cartesian orthogonal coordinates,

$$\nabla = \begin{pmatrix} \partial/\partial x \\ \partial/\partial y \\ \partial/\partial z \end{pmatrix}. \tag{3.18}$$

Since, for a free particle,

$$E = \frac{p^2}{2m}, \tag{3.19}$$

Schrödinger equation for the free particle immediately follows:

$$\frac{\partial \psi}{\partial t} = \frac{i\hbar}{2m} \nabla^2 \psi. \tag{3.20}$$

Applying the differential operators $i\hbar \partial/\partial t$ and $-i\hbar\nabla$ to the plane monochromatic wave:

$$\exp\left[i(\mathbf{k} \cdot \mathbf{r} - \omega t)\right],$$

we obtain the relationships:

$$E = \hbar\omega, \tag{3.21}$$

$$\mathbf{p} = \hbar\mathbf{k}.^{[1]} \tag{3.22}$$

Note that the time-independent Schrödinger equation for the free particle follows:

$$E\psi_{\mathbf{k}}(\mathbf{r}) = -\frac{\hbar^2}{2m} \nabla^2 \psi_{\mathbf{k}}(\mathbf{r}), \tag{3.23}$$

---

[1] Please note that, from Eq. (3.22), the de Broglie relationship connecting the momentum of a particle to its wavelength is as follows:

$$p = \frac{h}{\lambda}. \tag{3.24}$$

where

$$\psi_{\mathbf{k}}(\mathbf{r}) = \exp(i\mathbf{k} \cdot \mathbf{r}).$$ (3.25)

3.3 Continuity equation

Let us consider a conserved quantity such as, for example, the electric charge, which is distributed with density ρ inside a portion of space of volume V. Since the considered quantity is conserved, the flux of the current density $\mathbf{j}$ through the surface S surrounding the volume V must be equal and opposite to the volume integral of the partial derivative of the density with respect to time:

$$-\int_S \mathbf{j} \cdot d\mathbf{S} = \int_V \frac{\partial \rho}{\partial t} \, dV.$$ (3.26)

Let us apply now the divergence theorem:

$$\int_S \mathbf{j} \cdot d\mathbf{S} = \int_V \nabla \cdot \mathbf{j} \, dV.$$ (3.27)

Comparing the two latter equations, we obtain

$$\int_V \left(\nabla \cdot \mathbf{j} + \frac{\partial \rho}{\partial t} \right) dV = 0.$$ (3.28)

Since this result does not depend on the choice of the volume V, the integrand must be null. Thus, the continuity equation immediately follows:

$$\nabla \cdot \mathbf{j} + \frac{\partial \rho}{\partial t} = 0.$$ (3.29)

For the case of stationary conditions, we have

$$\frac{\partial \rho}{\partial t} = 0,$$ (3.30)

so that the continuity equation becomes

$$\nabla \cdot \mathbf{j} = 0.$$ (3.31)

Let us indicate now with ρ the probability density associated with the wave function ψ:

$$\rho = \psi^* \psi.$$ (3.32)

The probability current density **j** defined by

$$\mathbf{j} = \frac{i\hbar}{2m}(\psi \, \nabla \psi^* - \psi^* \, \nabla \psi) \tag{3.33}$$

satisfies the continuity equation [see Eq. (3.29)]. This result can be easily demonstrated using the Schrödinger equation (3.20) and its hermitian conjugate:

$$\frac{\partial \psi^*}{\partial t} = -\frac{i\hbar}{2m} \nabla^2 \psi^* \, . \tag{3.34}$$

Indeed

$$\nabla \cdot \mathbf{j} = \frac{i\hbar}{2m} [\nabla \cdot (\psi\nabla\psi^*) - \nabla \cdot (\psi^*\nabla\psi)] \, , \tag{3.35}$$

$$\nabla \cdot (\psi\nabla\psi^*) = \nabla\psi \cdot \nabla\psi^* + \psi\nabla^2\psi^* \, , \tag{3.36}$$

and

$$\nabla \cdot (\psi^*\nabla\psi) = \nabla\psi^* \cdot \nabla\psi + \psi^*\nabla^2\psi \, . \tag{3.37}$$

As a consequence,

$$\begin{aligned} \nabla \cdot \mathbf{j} &= \frac{i\hbar}{2m}[\nabla\psi \cdot \nabla\psi^* + \psi\nabla^2\psi^* - \nabla\psi^* \cdot \nabla\psi - \psi^*\nabla^2\psi] \\ &= \psi\frac{i\hbar}{2m}\nabla^2\psi^* - \psi^*\frac{i\hbar}{2m}\nabla^2\psi = -\psi\frac{\partial\psi^*}{\partial t} - \psi^*\frac{\partial\psi}{\partial t} \\ &= -\frac{\partial}{\partial t}(\psi^*\psi) = -\frac{\partial\rho}{\partial t} \, . \end{aligned}$$

3.4 Differential elastic scattering cross-section

The wave function $\psi_\mathbf{k}(\mathbf{r})$ of an incident electron beam moving in the direction of the z axis is the plane wave:

$$\psi_\mathbf{k}(\mathbf{r}) = \exp(i\mathbf{k} \cdot \mathbf{r}) = \exp(ikz) \, , \tag{3.38}$$

where

$$\mathbf{k} = \frac{m\mathbf{v}}{\hbar} \tag{3.39}$$

is the electron wavenumber, m the electron mass, and **v** the electron velocity. Please note that we are here dealing with an incident beam normalized to one electron per unit volume. Let us now consider the wave function $\psi_s(r, \vartheta, \varphi)$, which describes the electrons elastically scattered by an atom. After the collision, the scattered electrons can be described by the spherical wave [see Eq. (3.15)]:

$$\psi_s(r, \vartheta, \varphi) = f(\vartheta, \varphi) \frac{\exp(ikr)}{r},$$

where the function $f(\vartheta, \varphi)$ is known as the *scattering amplitude*. The scattering amplitude depends only on the polar angle ϑ and on the azimuthal angle φ, so all the radial dependence of the spherical wave is completely described by the factor $\exp(ikr)/r$. For large values of r,

$$-\frac{\hbar^2}{2m} \nabla^2 \psi_s = E \psi_s, \tag{3.40}$$

where

$$E = \frac{\hbar^2 k^2}{2m}. \tag{3.41}$$

In other words, the spherical wave function ψ_s satisfies, far from the scattering centre, the time-independent Schrödinger equation for the free electron. This can be easily demonstrated using Eqs. (3.6) and (3.7). In fact

$$\frac{\partial}{\partial r} \frac{\exp(ikr)}{r} = \left(ik - \frac{1}{r}\right) \frac{\exp(ikr)}{r}, \tag{3.42}$$

and

$$\frac{\partial^2}{\partial r^2} \frac{\exp(ikr)}{r} = \left(-k^2 - \frac{2ik}{r} + \frac{2}{r^2}\right) \frac{\exp(ikr)}{r}, \tag{3.43}$$

so that

$$\left(\frac{\partial^2}{\partial r^2} + \frac{2}{r} \frac{\partial}{\partial r}\right) \frac{\exp(ikr)}{r} = -k^2 \frac{\exp(ikr)}{r}. \tag{3.44}$$

As a consequence,

$$\left(\frac{\partial^2}{\partial r^2} + \frac{2}{r} \frac{\partial}{\partial r} + \frac{1}{r^2}\Lambda\right) f(\vartheta, \varphi) \frac{\exp(ikr)}{r}$$
$$= -k^2 f(\vartheta, \varphi) \frac{\exp(ikr)}{r} + \frac{1}{r^2}\Lambda f(\vartheta, \varphi) \frac{\exp(ikr)}{r}, \tag{3.45}$$

so, when $r \to \infty$,

$$\left(\frac{\partial^2}{\partial r^2} + \frac{2}{r} \frac{\partial}{\partial r} + \frac{1}{r^2}\Lambda\right) \psi_s = -k^2 \psi_s, \tag{3.46}$$

where we have neglected terms tending to zero faster than $1/r$. In general, the wave function describing the scattering process satisfies the Schrödinger equation in the presence of a potential $V(\mathbf{r})$:

$$\left[-\frac{\hbar^2}{2m}\nabla^2 + V(\mathbf{r})\right]\psi = E\psi.$$ (3.47)

Very far from the center of scattering the potential becomes negligible so that, to describe both the incident and the scattered particles, the wave function has to approach the sum of $\psi_{\mathbf{k}}$ and $\psi_{\mathbf{s}}$, i. e., it must satisfy the following boundary condition:

$$\psi(r, \vartheta, \varphi) \underset{r\to\infty}{\sim} \exp(ikz) + f(\vartheta, \varphi)\frac{\exp(ikr)}{r}.$$ (3.48)

Let us now consider the component j_r of the probability current density [see Eq. (3.33)] for particles moving away from the scattering center:

$$j_r = \frac{\hbar}{2mi}\left\{f^*\frac{\exp(-ikr)}{r}\frac{\partial}{\partial r}\left[f\frac{\exp(ikr)}{r}\right] - f\frac{\exp(ikr)}{r}\frac{\partial}{\partial r}\left[f^*\frac{\exp(-ikr)}{r}\right]\right\}.$$ (3.49)

Note that $f = f(\vartheta, \varphi)$ does not depend on r, so that

$$\frac{\partial}{\partial r}\left[f\frac{\exp(ikr)}{r}\right] = f\left[-\frac{\exp(ikr)}{r^2} + ik\frac{\exp(ikr)}{r}\right],$$

$$\frac{\partial}{\partial r}\left[f^*\frac{\exp(-ikr)}{r}\right] = f^*\left[-\frac{\exp(-ikr)}{r^2} - ik\frac{\exp(-ikr)}{r}\right].$$

As a consequence,

$$j_r = \frac{\hbar k}{m}\frac{|f(\vartheta, \varphi)|^2}{r^2}.$$ (3.50)

Since v is the velocity of the incident electrons, we have

$$v = \frac{\hbar k}{m},$$ (3.51)

and hence

$$j_r = v\frac{|f(\vartheta, \varphi)|^2}{r^2}.$$ (3.52)

Note that electron velocity v is identical with the incident flux of electrons, i. e., the number of particles per unit time and per unit area normalized to the density of particles, while the number of electrons emerging in the solid angle $d\Omega$ per unit time is $j_r r^2 d\Omega$. Consequently, the differential elastic scattering cross-section, i. e., the ratio between the number of electrons emerging in $d\Omega$ per unit time and per unit solid angle, and the incident flux v, is given by

$$\frac{d\sigma}{d\Omega} = \frac{j_r r^2}{v} = |f(\vartheta, \varphi)|^2 . \tag{3.53}$$

3.5 The radial equation

We know that the wave function describing the scattering process satisfies the Schrödinger equation

$$\left[-\frac{\hbar^2}{2m} \nabla^2 + V \right] \psi = E \psi ,$$

where $V = V(\mathbf{r})$ is the atomic potential energy. Once introduced, the quantity $U(\mathbf{r})$ defined as

$$U(\mathbf{r}) = \frac{2m}{\hbar^2} V(\mathbf{r}) , \tag{3.54}$$

the Schrödinger equation can be rewritten in the following simpler form:

$$(\nabla^2 + k^2) \psi = U\psi . \tag{3.55}$$

Let us now assume that our problem is spherically symmetric. Thus the potential energy V and, consequently, the quantity U depend only on the distance from origin r. So we can write that $V = V(r)$ and $U = U(r)$. Furthermore, from the observation that the wave function ψ must be axially symmetric around the electron beam, it follows that the former does not depend on the azimuthal angle φ, so that $\psi = \psi(r, \cos\theta)$. The set of the Legendre polynomials is complete, so that we can expand ψ as follows:

$$\psi(r, \cos\vartheta) = \sum_{l=0}^{\infty} C_l \frac{F_l(r)}{r} P_l(\cos\vartheta) , \tag{3.56}$$

where C_l are constants to be determined, $F_l(r)$ are functions of r to be determined, and $P_l(\cos\vartheta)$ are the Legendre polynomials. The functions $F_l(r)$ are called partial waves: in particular, $F_0(r)$ is the s-wave, $F_1(r)$ is the p-wave, and $F_2(r)$ is the d-wave. Now, we can write

$$(\nabla^2 + k^2 - U) \sum_{l=0}^{\infty} C_l \frac{F_l(r)}{r} P_l(\cos\vartheta) = 0 , \tag{3.57}$$

and, assuming that we can differentiate the sum term by term,

$$\sum_{l=0}^{\infty} (\nabla^2 + k^2 - U) C_l \frac{F_l(r)}{r} P_l(\cos\vartheta) = 0 . \tag{3.58}$$

The form of the Laplacian operator in spherical coordinates is expressed by Eq. (3.6), so

$$\sum_{l=0}^{\infty} C_l \left(\frac{\partial^2}{\partial r^2} + \frac{2}{r} \frac{\partial}{\partial r} + \frac{1}{r^2} \Lambda + k^2 - U \right) \frac{F_l(r)}{r} P_l(\cos \vartheta) = 0. \tag{3.59}$$

Keeping in mind that Legendre polynomials satisfy the following differential equation:

$$\left[(1 - u^2) \frac{d^2}{du^2} - 2u \frac{d}{du} + l(l+1) \right] P_l(u) = 0, \tag{3.60}$$

it can be easily demonstrated that they are eigenvalues of operator Λ defined by Eq. (3.7) with eigenvalues $-l(l+1)$. In fact,

$$\Lambda P_l(\cos \vartheta)$$

$$= \left(\frac{\partial^2}{\partial \vartheta^2} + \frac{\cos \vartheta}{\sin \vartheta} \frac{\partial}{\partial \vartheta} + \frac{1}{\sin^2 \vartheta} \frac{\partial^2}{\partial \varphi^2} \right) P_l(\cos \vartheta)$$

$$= \left(\frac{d^2}{d\vartheta^2} + \frac{\cos \vartheta}{\sin \vartheta} \frac{d}{d\vartheta} \right) P_l(\cos \vartheta)$$

$$= \left(\frac{1}{\sin \vartheta} \frac{d}{d\vartheta} \sin \theta \frac{d}{d\vartheta} \right) P_l(\cos \vartheta)$$

$$= \left(\frac{1}{\sin \vartheta} \frac{d}{d\vartheta} \sin^2 \vartheta \frac{1}{\sin \vartheta} \frac{d}{d\vartheta} \right) P_l(\cos \vartheta).$$

If $u = \cos \vartheta$, then $\sin^2 \vartheta = 1 - u^2$ and $du = -\sin \vartheta d\vartheta$, so that

$$\Lambda P_l(u) = \left[\frac{d}{du} (1 - u^2) \frac{d}{du} \right] P_l(u). \tag{3.61}$$

The last equation can be rearranged as follows:

$$\left[(1 - u^2) \frac{d^2}{du^2} - 2u \frac{d}{du} - \Lambda \right] P_l(u) = 0. \tag{3.62}$$

A comparison between Eq. (3.60) and Eq. (3.62) demonstrates that the eigenvalues of the Legendre polynomials are $-l(l+1)$:

$$\Lambda P_l(u) = -l(l+1) P_l(u). \tag{3.63}$$

We can thus rewrite Eq. (3.59) as follows:

$$\sum_{l=0}^{\infty} C_l \left[\frac{\partial^2}{\partial r^2} + \frac{2}{r} \frac{\partial}{\partial r} - \frac{l(l+1)}{r^2} + k^2 - U \right] \frac{F_l(r)}{r} P_l(\cos \vartheta) = 0. \tag{3.64}$$

The coefficients of all the Legendre polynomials $P_l(\cos \vartheta)$ in Eq. (3.64) can be equated to zero as the equation is valid for each value of ϑ, so that

$$\left[\frac{d^2}{dr^2} + \frac{2}{r} \frac{d}{dr} - \frac{l(l+1)}{r^2} + k^2 - U \right] \frac{F_l(r)}{r} = 0. \tag{3.65}$$

Since

$$\frac{d^2}{dr^2} \frac{F_l(r)}{r}$$

$$= \frac{d}{dr} \left(-\frac{F_l(r)}{r^2} + \frac{1}{r} \frac{dF_l(r)}{dr} \right)$$

$$= \frac{2}{r^3} F_l(r) - \frac{2}{r^2} \frac{dF_l(r)}{dr} + \frac{1}{r} \frac{d^2 F_l(r)}{dr^2}$$

and

$$\frac{2}{r} \frac{d}{dr} \frac{F_l(r)}{r}$$

$$= -\frac{2}{r^3} F_l(r) + \frac{2}{r^2} \frac{dF_l(r)}{dr},$$

thus

$$\left(\frac{d^2}{dr^2} + \frac{2}{r} \frac{d}{dr} \right) \frac{F_l(r)}{r} = \frac{1}{r} \frac{d^2 F_l(r)}{dr^2}.$$

As a consequence, Eq. (3.65) assumes the form

$$\left[\frac{d^2}{dr^2} - \frac{l(l+1)}{r^2} - U(r) + k^2 \right] F_l(r) = 0. \tag{3.66}$$

Equation (3.66) involves only r, the radial distance from the center of scattering. It is known as the radial equation. Note that, when $r \to \infty$, $U(r) \to 0$ and the radial equation assumes the form

$$\left(\frac{d^2}{dr^2} + k^2 \right) F_l(r) = 0, \tag{3.67}$$

then the solutions are simply given by sinusoidal functions:

$$F_l(r) \underset{r \to \infty}{\sim} A_l \sin(kr + \delta_l). \tag{3.68}$$

It is now convenient to define the *phase shifts* η_l so that

$$\eta_l = \delta_l + \frac{\pi l}{2}, \tag{3.69}$$

and therefore

$$F_l(r) \underset{r \to \infty}{\sim} A_l \sin\left(kr - \frac{\pi l}{2} + \eta_l\right). \tag{3.70}$$

Note that, since constants C_l are not yet specified, we can choose $A_l = 1$ for each $l = 0, \ldots, \infty$, and then write

$$F_l(r) \underset{r \to \infty}{\sim} \sin\left(kr - \frac{\pi l}{2} + \eta_l\right). \tag{3.71}$$

Let us now consider the case of a free particle, so that $U(r) = 0$. In this case we will indicate the partial waves with the symbols $F_l^0(r)$ and the corresponding phase shifts with the symbols η_l^0. The radial equations ($l = 0, \ldots, \infty$) assume the form:

$$\left[\frac{d^2}{dr^2} - \frac{l(l+1)}{r^2} + k^2\right] F_l^0(r) = 0. \tag{3.72}$$

The solutions of these equations are proportional to the regular spherical Bessel functions, $j_l(kr)$, multiplied by kr. To find the constant of proportionality and the phase shifts η_l^0, let us now still consider the limit corresponding to large values of r ($r \to \infty$):

$$krj_l(kr) \underset{r \to \infty}{\sim} \sin\left(kr - \frac{\pi l}{2}\right), \tag{3.73}$$

and compare these functions with the partial waves $F_l^0(r)$ in the same limit:

$$F_l^0(r) \underset{r \to \infty}{\sim} \sin\left(kr - \frac{\pi l}{2} + \eta_l^0\right). \tag{3.74}$$

The comparison demonstrates that the constant of proportionality is 1 and that the phase shifts η_l^0 corresponding to the free electrons are equal to zero, $\eta_l^0 = 0$, for each $l = 0, \ldots, \infty$. Thus, we have demonstrated that the potential $U(r)$ introduces a shift η_l in the phase of the scattered waves. If the potential is null, then the phase shift is null. Furthermore, if the potential is null, we have shown that the partial waves are simply given by the product of kr by the regular spherical Bessel functions $j_l(kr)$:

$$F_l^0(r) = kr \, j_l(kr). \tag{3.75}$$

3.6 Expansion of the plane wave

To proceed, we need to expand plane wave $\psi_\mathbf{k}$ on the complete set of Legendre polynomials:

$$\psi_\mathbf{k}(r, \cos\vartheta) = \exp(ikz) = \sum_{l=0}^{\infty} C_l^0 \frac{F_l^0(r)}{r} P_l(\cos\vartheta). \tag{3.76}$$

Taking into account Eq. (3.75), we obtain

$$\exp(ikz) = \exp(ikr\cos\vartheta) = \sum_{l=0}^{\infty}(C_l^0\,k)\,j_l(kr)\,P_l(\cos\vartheta) = \sum_{l=0}^{\infty}c_l\,j_l(kr)\,P_l(\cos\vartheta), \quad (3.77)$$

where $c_l = (C_l^0\,k)$ are constants to be determined. Let us introduce the two new variables $s \equiv kr$ and $u \equiv \cos\vartheta$, so that

$$\exp(isu) = \sum_{l=0}^{\infty}c_l j_l(s)P_l(u). \quad (3.78)$$

Now let us differentiate the previous equation with respect to the variable s, to obtain

$$iu\exp(isu) = \sum_{l=0}^{\infty}iuc_l j_l(s)P_l(u) = \sum_{l=0}^{\infty}c_l\frac{dj_l(s)}{ds}P_l(u). \quad (3.79)$$

The Legendre polynomials satisfy the equation:

$$P_l(u) = \frac{(l+1)P_{l+1}(u) + lP_{l-1}(u)}{u\,(2l+1)}. \quad (3.80)$$

Therefore,

$$
\begin{aligned}
iu\exp(isu) &= \sum_{l=0}^{\infty}i\,c_l\,j_l(s)\frac{(l+1)P_{l+1}(u) + lP_{l-1}(u)}{(2l+1)} \\
&= \sum_{l=1}^{\infty}iP_l(u)\left[\frac{l}{2l-1}c_{l-1}j_{l-1}(s)\right] + \sum_{l=-1}^{\infty}iP_l(u)\left[\frac{l+1}{2l+3}c_{l+1}j_{l+1}(s)\right] \\
&= \sum_{l=0}^{\infty}iP_l(u)\left[\frac{l}{2l-1}c_{l-1}j_{l-1}(s)\right] + \sum_{l=0}^{\infty}iP_l(u)\left[\frac{l+1}{2l+3}c_{l+1}j_{l+1}(s)\right] \\
&= \sum_{l=0}^{\infty}iP_l(u)\left[\frac{l}{2l-1}c_{l-1}j_{l-1}(s) + \frac{l+1}{2l+3}c_{l+1}j_{l+1}(s)\right]. \quad (3.81)
\end{aligned}
$$

Note that

$$s\,j_{l-1}(s) - (2l+1)\,j_l(s) + s\,j_{l+1}(s) = 0, \quad (3.82)$$

$$s\,j_{l-1}(s) - (l+1)\,j_l(s) - s\,\frac{dj_l(s)}{ds} = 0, \quad (3.83)$$

and, consequently, the derivatives of the regular spherical Bessel functions are given by:

$$\frac{dj_l(s)}{ds} = \frac{l}{2l+1}j_{l-1}(s) - \frac{l+1}{2l+1}j_{l+1}(s). \quad (3.84)$$

Therefore

$$iu \exp(isu) = \sum_{l=0}^{\infty} c_l P_l(u) \left[\frac{l}{2l+1} j_{l-1}(s) - \frac{l+1}{2l+1} j_{l+1}(s) \right]. \tag{3.85}$$

From Eqs. (3.81) and (3.85) we obtain

$$\sum_{l=0}^{\infty} P_l(u) \left[j_{l-1}(s) \, l \left(\frac{c_l}{2l+1} - \frac{ic_{l-1}}{2l-1} \right) \right.$$

$$\left. - j_{l+1}(s) \, (l+1) \left(\frac{c_l}{2l+1} + \frac{ic_{l+1}}{2l+3} \right) \right] = 0. \tag{3.86}$$

Since the Legendre polynomials $P_l(u)$ are linearly independent functions, the coefficient of each $P_l(u)$ in the sum can be equated to zero:

$$j_{l-1}(s) \, l \left(\frac{c_l}{2l+1} - \frac{ic_{l-1}}{2l-1} \right) = j_{l+1}(s) \, (l+1) \left(\frac{c_l}{2l+1} + \frac{ic_{l+1}}{2l+3} \right). \tag{3.87}$$

As the previous equation has to be verified for each value of s, we conclude that

$$\frac{1}{2l+1} c_l = \frac{i}{2l-1} c_{l-1}, \tag{3.88}$$

$$\frac{1}{2l+1} c_l = -\frac{i}{2l+3} c_{l+1}. \tag{3.89}$$

Eqs. (3.88) and (3.89) are equivalent. We just need to substitute l with $l-1$ in the second equation to find the first one. We need to know the value of the first coefficient of the set of coefficients c_l, i. e., c_0, to use Eq. (3.88) [or Eq. (3.89)] to calculate the others. Since

$$\exp(0) = 1 = \sum_{l=0}^{\infty} c_l j_l(0) P_l(\cos \vartheta), \tag{3.90}$$

$j_l(0) = 0$ for any $l \neq 0$, $j_0(0) = 1$, and $P_0(\cos \vartheta) = 1$, we conclude that $c_0 = 1$. Now, the recursive use of Eq. (3.88) allows us to obtain the values of the coefficients:

$$c_l = (2l+1) i^l, \tag{3.91}$$

and the expansion of the plane wave in Legendre polynomials:

$$\exp(ikr \cos \vartheta) = \exp(ikz) = \sum_{l=0}^{\infty} (2l+1) i^l j_l(kr) P_l(\cos \vartheta). \tag{3.92}$$

The same result can be obtained observing that, since

$$\int_{-1}^{1} P_m(u) P_n(u)\, du = \frac{2}{2n+1}\, \delta_{mn}\,,\tag{3.93}$$

we have

$$\int_{-1}^{+1} \sum_{n=0}^{\infty} c_n j_n(s) P_n(u) P_l(u)\, du = \frac{2}{2l+1}\, c_l j_l(s)\,,\tag{3.94}$$

and, as a consequence, taking into account Eq. (3.78),

$$\frac{2}{2l+1}\, c_l j_l(s) = \int_{-1}^{+1} \exp(isu) P_l(u)\, du\,.\tag{3.95}$$

Integrating by parts we obtain

$$\int_{-1}^{+1} \exp(isu) P_l(u)\, du = \left[\frac{\exp(isu)}{is} P_l(u) \right]_{-1}^{+1} - \int_{-1}^{+1} \frac{\exp(isu)}{is} P_l'(u)\, du$$

$$= \frac{\exp(is) P_l(1) - \exp(-is) P_l(-1)}{is} + h(s)\,,$$

where we have defined the function $h(s)$ as

$$h(s) \equiv - \int_{-1}^{+1} \frac{\exp(isu)}{is} P_l'(u)\, du\,.$$

Since $P_l(1) = 1$ and $P_l(-1) = (-1)^l$, we have

$$\frac{2 j_l(s) c_l}{2l+1} = \frac{2 i^l}{s} \sin\left(s - \frac{\pi l}{2} \right) + h(s)\,.\tag{3.96}$$

On the one hand, as $s \to \infty$, $h(s) \to 0$ faster than $1/s$. This can be easily verified by repeated integrations by parts. As $s \to \infty$, on the other hand,

$$j_l(s) \sim \frac{1}{s} \sin\left(s - \frac{\pi l}{2} \right)\,.\tag{3.97}$$

Comparing the last two equations in the limit $s \to \infty$, we obtain again Eq. (3.91) and, as a consequence, Eq. (3.92).

3.7 Scattering amplitude

The plane wave becomes, when $r \to \infty$,

$$\exp(ikz) \underset{r\to\infty}{\sim} \sum_{l=0}^{\infty} \frac{(2l+1)i^l}{kr} \sin\left(kr - \frac{\pi l}{2}\right) P_l(\cos\vartheta). \tag{3.98}$$

Taking into account Eq. (3.71), we see that, in the same limit, Eq. (3.56) assumes the form:

$$\psi(r,\cos\vartheta) \underset{r\to\infty}{\sim} \sum_{l=0}^{\infty} C_l \frac{1}{r} \sin\left(kr - \frac{\pi l}{2} + \eta_l\right) P_l(\cos\vartheta). \tag{3.99}$$

Consequently,

$$\psi(r,\cos\vartheta) - \exp(ikz) \underset{r\to\infty}{\sim} \frac{\exp(ikr)}{r} \mathcal{F}(\vartheta) + \frac{\exp(-ikr)}{r} \mathcal{G}(\vartheta), \tag{3.100}$$

where

$$\mathcal{F}(\vartheta) = \sum_{l=0}^{\infty} P_l(\cos\vartheta) \left\{ \frac{C_l}{2i} \exp\left[i\left(-\frac{\pi l}{2} + \eta_l\right)\right] - \frac{(2l+1)i^l}{2ik} \exp\left(-i\frac{\pi l}{2}\right) \right\}, \tag{3.101}$$

and

$$\mathcal{G}(\vartheta) = \sum_{l=0}^{\infty} P_l(\cos\vartheta) \left\{ -\frac{C_l}{2i} \exp\left[i\left(\frac{\pi l}{2} - \eta_l\right)\right] + \frac{(2l+1)i^l}{2ik} \exp\left(i\frac{\pi l}{2}\right) \right\}. \tag{3.102}$$

On the other hand, from Eq. (3.48), we know that

$$\psi(r,\cos\vartheta) - \exp(ikz) \underset{r\to\infty}{\sim} \frac{\exp(ikr)}{r} f(\vartheta). \tag{3.103}$$

The comparison between Eqs. (3.100) and (3.103) enables us to conclude, in particular, that

$$\mathcal{G}(\vartheta) = 0. \tag{3.104}$$

Therefore, from Eq. (3.102), it follows that

$$\frac{C_l}{2i} \exp\left[i\left(\frac{\pi l}{2} - \eta_l\right)\right] = \frac{(2l+1)i^l}{2ik} \exp\left(i\frac{\pi l}{2}\right). \tag{3.105}$$

This last equation enables us to express the coefficients C_l as a function of the phase shifts η_l, for $l = 0,\ldots,\infty$:

$$C_l = \frac{(2l+1)i^l \exp(i\eta_l)}{k}. \tag{3.106}$$

Another consequence of the comparison between Eqs. (3.100) and (3.103) is that

$$\mathcal{F}(\vartheta) = f(\vartheta).$$ (3.107)

Therefore, taking also into account Eq. (3.106), we can express the scattering amplitude as the following expansion on the complete set of Legendre's polynomials:

$$f(\vartheta) = \frac{1}{2ik} \sum_{l=0}^{\infty} (2l+1) \left[\exp(2i\eta_l) - 1\right] P_l(\cos\vartheta).$$ (3.108)

This expansion can also be expressed as

$$f(\vartheta) = \frac{1}{k} \sum_{l=0}^{\infty} (2l+1) \exp(i\eta_l) \sin\eta_l P_l(\cos\vartheta).$$ (3.109)

3.8 Total elastic scattering cross-section and optical theorem

The total elastic scattering cross-section is given by

$$\sigma = \int \frac{d\sigma}{d\Omega} d\Omega = 2\pi \int_0^{\pi} \frac{d\sigma}{d\Omega} \sin\vartheta \, d\vartheta,$$ (3.110)

where the differential elastic scattering cross-section $d\sigma/d\Omega$ is the squared modulus of the scattering amplitude:

$$\frac{d\sigma}{d\Omega} = \frac{1}{k^2} \sum_{m=0}^{\infty} \sum_{n=0}^{\infty} (2m+1)(2n+1) \exp\left[i(\eta_m - \eta_n)\right] \sin\eta_m \sin\eta_n P_m(\cos\vartheta) P_n(\cos\vartheta).$$ (3.111)

Using Eqs. (3.93), and (3.110), (3.111) we obtain

$$\sigma = \frac{4\pi}{k^2} \sum_{l=0}^{\infty} (2l+1) \sin^2\eta_l.$$ (3.112)

Note that the imaginary part of the forward scattering amplitude $f(0)$ is given by

$$\mathrm{Im}\, f(0) = \frac{1}{k} \sum_{l=0}^{\infty} (2l+1) \sin^2\eta_l,$$ (3.113)

so that

$$\sigma = \frac{4\pi}{k} \mathrm{Im}\, f(0).$$ (3.114)

This result is known as the "optical theorem".

3.9 The first Born approximation

If electron energy is high enough, another approach allows to describe the electron–atom elastic scattering, i. e., the scattering of a beam of electrons by a central potential $V(r)$. This approach is based on the first Born approximation. The first Born approximation is quite accurate if

$$E \gg \frac{e^2}{2a_0} Z^2 . \tag{3.115}$$

The Schrödinger equation,

$$(\nabla^2 + k^2)\, \psi(\mathbf{r}) = \frac{2m}{\hbar^2} V(\mathbf{r})\psi(\mathbf{r}), \tag{3.116}$$

is equivalent to the following integral equation:

$$\psi(\mathbf{r}) = \exp(ikz) + \frac{2m}{\hbar^2} \int d^3 r'\, g(\mathbf{r}, \mathbf{r}') V(\mathbf{r}')\psi(\mathbf{r}'), \tag{3.117}$$

where $g(\mathbf{r}, \mathbf{r}')$ is the Green function [see Eq. (2.65)].

To demonstrate Eq. (3.117), let us apply the operator $\nabla^2 + k^2$ to the function $\psi(\mathbf{r})$ defined by the integral equation (3.117):

$$\begin{aligned}
(\nabla^2 + k^2)\, \psi(\mathbf{r}) &= (\nabla^2 + k^2)\, \exp(ikz) \\
&+ \frac{2m}{\hbar^2} \int d^3 r'\, (\nabla^2 + k^2)\, g(\mathbf{r}, \mathbf{r}')\, V(\mathbf{r}')\, \psi(\mathbf{r}').
\end{aligned} \tag{3.118}$$

The application of the operator ∇^2 to the plane wave gives

$$\nabla^2 \exp(ikz) = \frac{\partial^2}{\partial z^2} \exp(ikz) = -k^2 \exp(ikz), \tag{3.119}$$

so that

$$(\nabla^2 + k^2)\exp(ikz) = 0. \tag{3.120}$$

Therefore,

$$\begin{aligned}
(\nabla^2 + k^2)\psi(\mathbf{r}) &= \frac{2m}{\hbar^2} \int d^3 r'\, (\nabla^2 + k^2)\, g(\mathbf{r}, \mathbf{r}')\, V(\mathbf{r}')\psi(\mathbf{r}') \\
&= \frac{2m}{\hbar^2} \int d^3 r'\, \delta(\mathbf{r} - \mathbf{r}') V(\mathbf{r}')\psi(\mathbf{r}') = \frac{2m}{\hbar^2} V(\mathbf{r})\, \psi(\mathbf{r}).
\end{aligned} \tag{3.121}$$

To obtain the scattering amplitude, please note that

$$|\mathbf{r} - \mathbf{r}'| = \sqrt{r^2 - 2\mathbf{r}\cdot\mathbf{r}' + r'^2}$$

$$= r\sqrt{1 - \frac{2\hat{\mathbf{r}}\cdot\mathbf{r}'}{r} + \frac{r'^2}{r^2}} \sim r\left(1 - \frac{\hat{\mathbf{r}}\cdot\mathbf{r}'}{r} + \mathcal{O}\left(\frac{1}{r^2}\right)\right), \tag{3.122}$$

where

$$\hat{\mathbf{r}} = \frac{\mathbf{r}}{r}. \tag{3.123}$$

Let us introduce $\mathcal{K}$, the wave number in the direction of the outgoing unit vector $\hat{\mathbf{r}}$:

$$\mathcal{K} = k\hat{\mathbf{r}}. \tag{3.124}$$

The Green function has thus the following asymptotic behavior:

$$g(\mathbf{r},\mathbf{r}') \underset{r\to\infty}{\sim} -\frac{\exp(ikr - i\mathcal{K}\cdot\mathbf{r}')}{4\pi r}. \tag{3.125}$$

Let us now introduce Eq. (3.125) into the integral equation (3.117), so that

$$\psi(\mathbf{r}) \underset{r\to\infty}{\sim} \exp(i\mathcal{K}z) - \frac{2m}{\hbar^2}\int d^3r' \frac{\exp(ikr - i\mathcal{K}\cdot\mathbf{r}')}{4\pi r} V(\mathbf{r}')\psi(\mathbf{r}'). \tag{3.126}$$

From

$$\int d^3r' \frac{\exp(ikr - i\mathcal{K}\cdot\mathbf{r}')}{4\pi r} V(\mathbf{r}')\psi(\mathbf{r}') = \frac{\exp(ikr)}{r}\int d^3r' \frac{\exp(-i\mathcal{K}\cdot\mathbf{r}')}{4\pi} V(\mathbf{r}')\psi(\mathbf{r}'),$$

we see that, if the scattering amplitude is given by

$$f(\vartheta,\varphi) = -\frac{m}{2\pi\hbar^2}\int d^3r \exp(-i\mathcal{K}\cdot\mathbf{r}) V(\mathbf{r})\psi(\mathbf{r}), \tag{3.127}$$

then the boundary conditions, Eq. (3.48), are satisfied. In the first Born approximation, due to the high ratio between the electron kinetic energy and the atomic potential energy, we have

$$\psi(\mathbf{r}) \approx \exp(ikz) = \exp(i\mathbf{k}\cdot\mathbf{r}), \tag{3.128}$$

and the equation (3.127) becomes

$$f(\vartheta,\varphi) = -\frac{m}{2\pi\hbar^2}\int d^3r \exp(-i\mathcal{K}\cdot\mathbf{r}) V(\mathbf{r}) \exp(i\mathbf{k}\cdot\mathbf{r}). \tag{3.129}$$

Indicating with $\hbar\mathbf{q}$ the momentum lost by the incident electron,

$$\hbar\mathbf{q} = \hbar(\mathbf{k} - \mathcal{K}), \tag{3.130}$$

we can write that

$$f(\vartheta, \varphi) = -\frac{m}{2\pi\hbar^2} \int d^3r \, \exp(i\mathbf{q} \cdot \mathbf{r}) \, V(\mathbf{r}) \,. \tag{3.131}$$

Now, since we are interested in a central potential,

$$V(\mathbf{r}) = V(r)$$

and

$$f(\vartheta, \varphi) = f(\vartheta)$$

$$= -\frac{m}{2\pi\hbar^2} \int_0^{2\pi} d\phi \int_0^{\pi} \sin\theta \, d\theta \int_0^{\infty} r^2 \, dr \, \exp(iqr\cos\theta) \, V(r)$$

$$= -\frac{2m}{\hbar^2 q} \int_0^{\infty} \sin(qr) \, V(r) \, r \, dr \,. \tag{3.132}$$

3.10 Rutherford elastic scattering cross-section

We are interested now in the calculation of Rutherford elastic scattering cross-section. This can be obtained using the first Born approximation with the following potential (Wentzel-like atomic potential):

$$V(r) = -\frac{Ze^2}{r} \exp\left(-\frac{r}{a}\right). \tag{3.133}$$

Please note that the a parameter can be approximated by

$$a = \frac{a_0}{Z^{1/3}} \,, \tag{3.134}$$

where $a_0 = \hbar^2/me^2$ is the Bohr radius. Using the first Born approximation, the scattering amplitude is given by

$$f(\vartheta) = \frac{2m}{\hbar^2} \frac{Ze^2}{q} \int_0^{\infty} \sin(qr) \, \exp\left(-\frac{r}{a}\right) dr \,. \tag{3.135}$$

From

$$\int_0^{\infty} \sin(qr) \, \exp\left(-\frac{r}{a}\right) dr = \frac{q}{q^2 + (1/a)^2} \,, \tag{3.136}$$

we obtain

$$\frac{d\sigma}{d\Omega} = |f(\vartheta)|^2 = \frac{4m^2}{\hbar^4} \frac{Z^2 e^4}{[q^2 + (1/a)^2]^2} .$$

(3.137)

Keeping in mind that $|\mathbf{k}| = |\mathcal{K}|$ and $\mathbf{q} = \mathbf{k} - \mathcal{K}$, we have

$$q^2 = (\mathbf{k} - \mathcal{K}) \cdot (\mathbf{k} - \mathcal{K}) = \mathbf{k}^2 + \mathcal{K}^2 - 2k\mathcal{K} \cos \vartheta = 2\mathbf{k}^2 (1 - \cos \vartheta),$$

(3.138)

where ϑ is the scattering angle. As the electron kinetic energy is given by

$$E = \frac{\hbar^2 \mathbf{k}^2}{2m},$$

(3.139)

the differential elastic scattering cross-section for the collision of an electron beam with a Wentzel-like atomic potential is given, in the first Born approximation, by

$$\frac{d\sigma}{d\Omega} = \frac{Z^2 e^4}{4E^2} \frac{1}{(1 - \cos \vartheta + \alpha)^2},$$

(3.140)

where

$$\alpha = \frac{1}{2k^2 a^2} = \frac{me^4 \pi^2}{\hbar^2} \frac{Z^{2/3}}{E}.$$

(3.141)

The well-known classical Rutherford formula,

$$\frac{d\sigma}{d\Omega} = \frac{Z^2 e^4}{4E^2} \frac{1}{(1 - \cos \vartheta)^2},$$

(3.142)

is the limit of Eq. (3.140) as $\alpha \to 0$. Equation (3.142) describes the elastic scattering of a beam of electrons by a pure Coulomb potential.

4 Spectrum of angular momentum and spin

The behavior of a complex atom in a magnetic field, as described by the Zeeman effect and by the Stern and Gerlach experiment, supports the hypothesis for the existence of the intrinsic angular momentum, or spin, of an electron. In this chapter we will briefly introduce the Pauli theory, i. e., the nonrelativistic theory of spin-1/2 particles [5, 14, 18].

4.1 Spectrum of angular momentum

We know that, in quantum mechanics, differential operators act on wave functions that are square-integrable complex functions in a Hilbert space. Let us take into account the components of the electron orbital angular momentum $\mathbf{L} = \mathbf{r} \times \mathbf{p}$, where we have indicated with $\mathbf{r}$ the electron position and with $\mathbf{p}$ the electron momentum. Using the definition of $\mathbf{L}$ and the correspondence rule expressed by Eq. (3.17):

$$\mathbf{p} \rightarrow -i\hbar\nabla,$$

it is easy to see that

$$[L_x, L_y] = i\hbar L_z, \tag{4.1}$$
$$[L_y, L_z] = i\hbar L_x, \tag{4.2}$$
$$[L_z, L_x] = i\hbar L_y, \tag{4.3}$$

where $[A, B]$ is the commutator of the operators A and B.[1] Let us consider, for example, the commutator $[L_x, L_y]$:

$$L_x L_y - L_y L_x$$
$$= -\hbar^2 \left[\left(y\frac{\partial}{\partial z} - z\frac{\partial}{\partial y} \right)\left(z\frac{\partial}{\partial x} - x\frac{\partial}{\partial z} \right) - \left(z\frac{\partial}{\partial x} - x\frac{\partial}{\partial z} \right)\left(y\frac{\partial}{\partial z} - z\frac{\partial}{\partial y} \right) \right]$$
$$= -\hbar^2 \left(y\frac{\partial}{\partial x} - x\frac{\partial}{\partial y} \right) = \frac{\hbar^2}{-i\hbar} L_z = i\hbar L_z.$$

Let us now consider a more general case. We will say that a linear operator $\mathbf{J}$ is an angular momentum, orbital or intrinsic, if its components obey the commutation rules:

$$[J_x, J_y] = i\hbar J_z, \tag{4.5}$$

---

1 The commutator is defined for any pair of linear operators A and B such that

$$[A, B] = AB - BA. \tag{4.4}$$

https://doi.org/10.1515/9783110675375-004

$$[J_y, J_z] = i\hbar J_x, \tag{4.6}$$
$$[J_z, J_x] = i\hbar J_y. \tag{4.7}$$

Since[2]

$$J_z J_x^2 - J_x^2 J_z = J_z J_x J_x - J_x^2 J_z = J_x J_z J_x - [J_x, J_z]J_x - J_x^2 J_z$$
$$= J_x^2 J_z - J_x[J_x, J_z] - [J_x, J_z]J_x - J_x^2 J_z$$
$$= -\{J_x, [J_x, J_z]\} = -\{J_x, -i\hbar J_y\} = i\hbar\{J_x, J_y\},$$
$$J_z J_y^2 - J_y^2 J_z = -\{J_y, [J_y, J_z]\} = -\{J_y, i\hbar J_x\}$$
$$= -i\hbar\{J_y, J_x\} = -i\hbar\{J_x, J_y\},$$

and

$$J_z J_z^2 - J_z^2 J_z = 0,$$

the two operators J_z and $\mathbf{J}^2$ commute:

$$[J_z, \mathbf{J}^2] = 0. \tag{4.9}$$

This means that J_z and $\mathbf{J}^2$ possess at least one basis of eigenvectors in common and, consequently, the physical quantities represented by these two operators may be measured together and simultaneously with arbitrary precision. Using the Dirac notation, let us indicate with $|jm\rangle$ a basis of orthonormal eigenvectors common to J_z and $\mathbf{J}^2$. The eigenvalues $\hbar m$ and $\hbar^2 j(j+1)$ correspond, respectively, to J_z and $\mathbf{J}^2$. Thus,[3]

$$J_z|jm\rangle = \hbar m|jm\rangle, \tag{4.10}$$
$$\mathbf{J}^2|jm\rangle = \hbar^2 j(j+1)|jm\rangle. \tag{4.11}$$

Let us now define two new operators, whose meaning will soon become clear. They are the operators J_- and J_+, given by

$$J_\pm = J_x \pm iJ_y. \tag{4.12}$$

J_- and J_+ have the following properties:

$$J_\pm^\dagger = J_\mp, \tag{4.13}$$

---

2 The anticommutator is defined for any pair of linear operators A and B such that

$$\{A, B\} = AB + BA. \tag{4.8}$$

3 Please note that here we use the symbol m to indicate the eigenvalues of J_z divided by $\hbar$. The context always enables avoiding confusion with the electron mass, also indicated with m elsewhere.

$$[J_z, J_\pm] = \pm\hbar J_\pm , \tag{4.14}$$

$$J_+J_- = J_x^2 + J_y^2 + \hbar J_z , \tag{4.15}$$

$$J_-J_+ = J_x^2 + J_y^2 - \hbar J_z , \tag{4.16}$$

$$[J_+, J_-] = 2\hbar J_z , \tag{4.17}$$

$$\{J_+, J_-\} = 2(J_x^2 + J_y^2). \tag{4.18}$$

These properties can be easily demonstrated. The first one is evident, as

$$J_\pm^\dagger = (J_x \pm iJ_y)^\dagger = (J_x \mp iJ_y) = J_\mp .$$

Let us now compute the commutator of J_z and $J_\pm$:

$$
\begin{aligned}
[J_z, J_\pm] &= [J_z, J_x] \pm i[J_z, J_y] \\
&= i\hbar J_y \pm i(-i\hbar J_x) = \hbar(iJ_y \pm J_x) \\
&= \pm\hbar(J_x \pm iJ_y) = \pm\hbar J_\pm .
\end{aligned}
$$

We can also easily calculate the commutator and the anticommutator of J_+ and J_+ just by observing that

$$
\begin{aligned}
J_+J_- &= J_x^2 + J_y^2 + iJ_yJ_x - iJ_xJ_y \\
&= J_x^2 + J_y^2 - i[J_x, J_y] = J_x^2 + J_y^2 - i(i\hbar J_z) \\
&= J_x^2 + J_y^2 + \hbar J_z
\end{aligned}
$$

and

$$
\begin{aligned}
J_-J_+ &= J_x^2 + J_y^2 - iJ_yJ_x + iJ_xJ_y \\
&= J_x^2 + J_y^2 + i[J_x, J_y] = J_x^2 + J_y^2 + i(i\hbar J_z) \\
&= J_x^2 + J_y^2 - \hbar J_z .
\end{aligned}
$$

Eqs. (4.17) and (4.18) immediately follow from the last two equations. Consequently,

$$\mathbf{J}^2 = J_z^2 + \frac{1}{2}(J_+J_- + J_-J_+). \tag{4.19}$$

Now we wish to calculate the norms of the vectors $J_-|jm\rangle$ and $J_+|jm\rangle$. Since

$$
\begin{aligned}
\langle jm|J_-^\dagger J_-|jm\rangle &= \langle jm|J_+J_-|jm\rangle \\
&= \langle jm|J_x^2 + J_y^2 + \hbar J_z|jm\rangle = \langle jm|\mathbf{J}^2 - J_z^2 + \hbar J_z|jm\rangle \\
&= \langle jm|jm\rangle \left[\hbar^2 j(j+1) - \hbar^2 m^2 + \hbar^2 m\right] = \hbar^2\left[j(j+1) - m(m-1)\right],
\end{aligned}
$$

we have

$$\langle jm|J_-^\dagger J_-|jm\rangle = \hbar^2[j(j+1) - m(m-1)] \geq 0, \tag{4.20}$$

and, similarly,

$$\langle jm|J_+^\dagger J_+|jm\rangle = \hbar^2[j(j+1) - m(m+1)] \geq 0. \tag{4.21}$$

Since they are positive or null, we have

$$m^2 - m - j(j+1) \leq 0 \tag{4.22}$$

and

$$m^2 + m - j(j+1) \leq 0, \tag{4.23}$$

so that

$$-j \leq m \leq j. \tag{4.24}$$

Furthermore, it can be easily demonstrated that $\mathbf{J}^2$ commutes with $J_\pm$:

$$[\mathbf{J}^2, J_\pm] = 0, \tag{4.25}$$

so that

$$\mathbf{J}^2(J_\pm|jm\rangle) = J_\pm \mathbf{J}^2|jm\rangle = \hbar^2 j(j+1)(J_\pm|jm\rangle). \tag{4.26}$$

This means that $J_\pm|jm\rangle$ are eigenvectors of $\mathbf{J}^2$ with eigenvalues $\hbar^2 j(j+1)$. From (4.14) it follows that

$$J_z(J_\pm|jm\rangle) = J_\pm J_z|jm\rangle \pm \hbar J_\pm|jm\rangle = \hbar(m \pm 1)(J_\pm|jm\rangle), \tag{4.27}$$

so that $J_\pm|jm\rangle$ are eigenvectors of J_z with eigenvalues $\hbar(m \pm 1)$. Comparing the previous equation with

$$J_z|jm \pm 1\rangle = \hbar(m \pm 1)|jm \pm 1\rangle, \tag{4.28}$$

we see that the eigenvectors $J_\pm|jm\rangle$ are proportional to $|jm \pm 1\rangle$:

$$J_\pm|jm\rangle = A_\pm|jm \pm 1\rangle, \tag{4.29}$$

where $A_\pm$ are two real and positive constants. From Eq. (4.20) it follows that

$$\langle jm - 1|A_-^* A_-|jm - 1\rangle$$
$$= |A_-|^2 \langle jm - 1|jm - 1\rangle$$
$$= A_-^2 = \hbar^2[j(j + 1) - m(m - 1)], \qquad (4.30)$$

so that

$$A_- = \hbar\sqrt{j(j + 1) - m(m - 1)}. \qquad (4.31)$$

Similarly, from Eq. (4.21) it follows that

$$A_+ = \hbar\sqrt{j(j + 1) - m(m + 1)}. \qquad (4.32)$$

Consequently,

$$J_\pm|jm\rangle = \hbar\sqrt{j(j + 1) - m(m \pm 1)}|jm \pm 1\rangle. \qquad (4.33)$$

Let us now indicate with $m_{\min}$ and $m_{\max}$ the minimum and the maximum values of m, respectively.[4] Let us now consider the following procedure:

$$J_-|jj\rangle = A_-|jj - 1\rangle,$$
$$J_-^2|jj\rangle = J_-A_-|jj - 1\rangle = A_-^2|jj - 2\rangle,$$
$$\dots$$
$$J_-^k|jj\rangle = A_-^k|j - j\rangle.$$

------

4 Since
$$J_-|jm_{\min}\rangle = 0,$$
we have
$$j(j + 1) - m_{\min}(m_{\min} - 1) = 0,$$
so that, taking into account Eq. (4.24),
$$m_{\min} = -j.$$
Similarly, from
$$J_+|jm_{\max}\rangle = 0,$$
we obtain
$$j(j + 1) - m_{\max}(m_{\max} + 1) = 0,$$
so that, still taking into account Eq. (4.24),
$$m_{\max} = j.$$

Here k is, for any given j, the integer number of steps from $m = m_{max} = j$ to $m = m_{min} = -j$. As a consequence, we can write that $j - k = -j$, so that $j = k/2$. In other words, by going from $m = m_{max} = j$ to $m = m_{min} = -j$ using the operator J_-, we can see that j must be either an integer non-negative number

$$j = 0,1,2,\dots,\infty, \tag{4.34}$$

or a half-integer non-negative number

$$j = \frac{1}{2}, \frac{3}{2}, \frac{5}{2}, \dots, \infty. \tag{4.35}$$

The only possible values of m, for any given j, are the $2j + 1$ numbers:

$$-j, -j+1, \dots, j-1, j. \tag{4.36}$$

In short, if $\hbar^2 j(j + 1)$ and $\hbar m$ are the eigenvalues of $\mathbf{J}^2$ and J_z, respectively, then j must be an integer or half-integer non-negative quantity and the values of m, for any given j, are the integer or half-integer numbers: $-j, -j+1, \dots, j-1, j$.

4.2 Spin

The intrinsic angular momentum, or spin, of an electron corresponds to the case $j = 1/2$. In this case, then, $j(j + 1) = 3/4$, and m can assume only the two values $m_{min} = -j = -1/2$ and $m_{max} = +j = +1/2$. Let us introduce the notation:

$$|-\rangle \equiv |1/2 - 1/2\rangle, \tag{4.37}$$
$$|+\rangle \equiv |1/2 + 1/2\rangle, \tag{4.38}$$

and indicate the spin operator by using the symbol **S**. The two-dimensional spin eigenspace is generated by the two eigenvectors $|-\rangle$ and $|+\rangle$. They have the following properties:

$$\langle +|-\rangle = \langle -|+\rangle = 0, \tag{4.39}$$
$$\langle -|-\rangle = \langle +|+\rangle = 1, \tag{4.40}$$
$$\mathbf{S}^2|-\rangle = \frac{3\hbar^2}{4}|-\rangle, \tag{4.41}$$
$$\mathbf{S}^2|+\rangle = \frac{3\hbar^2}{4}|+\rangle, \tag{4.42}$$
$$S_z|-\rangle = -\frac{\hbar}{2}|-\rangle, \tag{4.43}$$
$$S_z|+\rangle = \frac{\hbar}{2}|+\rangle. \tag{4.44}$$

In general, the spin-1/2 state $|\alpha\rangle$ will be a linear superposition of the two basic eigenvectors $|-\rangle$ and $|+\rangle$:

$$|\alpha\rangle = A|+\rangle + B|-\rangle. \tag{4.45}$$

In the previous equation, the coefficients A and B are complex numbers such that $|A|^2$ is the probability of finding the electron in the state of "spin up" along the z axis and $|B|^2$ is the probability of finding the electron in the "spin-down" state. The condition of normalization of $|\alpha\rangle$ entails that

$$|A|^2 + |B|^2 = 1. \tag{4.46}$$

Indeed,

$$
\begin{aligned}
1 = \langle\alpha|\alpha\rangle &= ((\langle+|A^\dagger + \langle-|B^\dagger)(A|+\rangle + B|-\rangle)) \\
&= |A|^2\langle+|+\rangle + |B|^2\langle-|-\rangle + A^\dagger B\langle+|-\rangle + B^\dagger A\langle-|+\rangle \\
&= |A|^2 + |B|^2.
\end{aligned}
$$

The matrix elements of S_+ and S_- can be easily calculated taking into account that

$$S_-|-\rangle = 0, \tag{4.47}$$
$$S_-|+\rangle = \hbar|-\rangle, \tag{4.48}$$
$$S_+|-\rangle = \hbar|+\rangle, \tag{4.49}$$
$$S_+|+\rangle = 0. \tag{4.50}$$

Let us calculate, for example, $S_-|+\rangle$ and $S_+|-\rangle$:

$$S_-|+\rangle = \hbar\sqrt{j(j+1) - m(m-1)}|-\rangle = \hbar\sqrt{3/4 + 1/4}|-\rangle = \hbar|-\rangle,$$
$$S_+|-\rangle = \hbar\sqrt{j(j+1) - m(m+1)}|+\rangle = \hbar\sqrt{3/4 + 1/4}|+\rangle = \hbar|+\rangle.$$

By using Eqs. (4.47), (4.48), (4.49), and (4.50), we obtain

$$\langle+|S_+|+\rangle = 0,$$
$$\langle+|S_+|-\rangle = \hbar\langle+|+\rangle = \hbar,$$
$$\langle-|S_+|-\rangle = \hbar\langle-|+\rangle = 0,$$
$$\langle-|S_+|+\rangle = 0,$$
$$\langle+|S_-|+\rangle = \hbar\langle+|-\rangle = 0$$
$$\langle+|S_-|-\rangle = 0,$$
$$\langle-|S_-|-\rangle = 0,$$
$$\langle-|S_-|+\rangle = \hbar\langle-|-\rangle = \hbar.$$

As a consequence, S_+ and S_- can be represented as:

$$S_+ = \hbar \begin{pmatrix} 0 & 1 \\ 0 & 0 \end{pmatrix}, \tag{4.51}$$

$$S_- = \hbar \begin{pmatrix} 0 & 0 \\ 1 & 0 \end{pmatrix}. \tag{4.52}$$

Eqs. (4.47), (4.48), (4.49), and (4.50) can be also easily verified using this representation. Let us consider, for example, Eq. (4.48):

$$S_-|+\rangle = \hbar \begin{pmatrix} 0 & 0 \\ 1 & 0 \end{pmatrix} \begin{pmatrix} 1 \\ 0 \end{pmatrix} = \hbar \begin{pmatrix} 0 \\ 1 \end{pmatrix} = \hbar|-\rangle. \tag{4.53}$$

The so-called Pauli matrices σ_x, σ_y, and σ_z are defined as

$$\mathbf{S} = \frac{\hbar}{2}\boldsymbol{\sigma}, \tag{4.54}$$

where

$$\boldsymbol{\sigma} = \begin{pmatrix} \sigma_x \\ \sigma_y \\ \sigma_z \end{pmatrix}, \tag{4.55}$$

and

$$S_+ = S_x + iS_y, \tag{4.56}$$

so we have

$$S_x = \frac{1}{2}(S_- + S_+), \tag{4.57}$$

$$S_y = \frac{i}{2}(S_- - S_+). \tag{4.58}$$

It follows that the representation of the Pauli matrices in the basis $\{|-\rangle, |+\rangle\}$ is given by

$$\sigma_x = \begin{pmatrix} 0 & 1 \\ 1 & 0 \end{pmatrix}, \tag{4.59}$$

$$\sigma_y = \begin{pmatrix} 0 & -i \\ i & 0 \end{pmatrix}, \tag{4.60}$$

$$\sigma_z = \begin{pmatrix} 1 & 0 \\ 0 & -1 \end{pmatrix}. \tag{4.61}$$

Indeed,

$$S_x = \frac{\hbar}{2}\begin{pmatrix} 0 & 0 \\ 1 & 0 \end{pmatrix} + \frac{\hbar}{2}\begin{pmatrix} 0 & 1 \\ 0 & 0 \end{pmatrix} = \frac{\hbar}{2}\begin{pmatrix} 0 & 1 \\ 1 & 0 \end{pmatrix}, \tag{4.62}$$

$$S_y = \frac{i}{2}\hbar\begin{pmatrix} 0 & 0 \\ 1 & 0 \end{pmatrix} - \frac{i}{2}\hbar\begin{pmatrix} 0 & 1 \\ 0 & 0 \end{pmatrix} = \frac{\hbar}{2}\begin{pmatrix} 0 & -i \\ i & 0 \end{pmatrix}. \tag{4.63}$$

Please note that the representation of S_z can be directly obtained by calculating the matrix elements as follows:

$$\langle +|S_z|+\rangle = \frac{\hbar}{2},$$

$$\langle +|S_z|-\rangle = 0,$$

$$\langle -|S_z|-\rangle = -\frac{\hbar}{2},$$

$$\langle -|S_z|+\rangle = 0.$$

Hence,

$$S_z = \frac{\hbar}{2}\begin{pmatrix} 1 & 0 \\ 0 & -1 \end{pmatrix}. \tag{4.64}$$

The Pauli matrices are Hermitian operators and have the following properties:

$$\sigma_x^2 = \sigma_y^2 = \sigma_z^2 = 1, \tag{4.65}$$

$$\sigma_x\sigma_y\sigma_z = i, \tag{4.66}$$

$$\sigma_x\sigma_y - \sigma_y\sigma_x = 2i\sigma_z, \tag{4.67}$$

$$\sigma_z\sigma_x - \sigma_x\sigma_z = 2i\sigma_y, \tag{4.68}$$

$$\sigma_y\sigma_z - \sigma_z\sigma_y = 2i\sigma_x, \tag{4.69}$$

$$\sigma_x\sigma_y + \sigma_y\sigma_x = 0, \tag{4.70}$$

$$\sigma_z\sigma_x + \sigma_x\sigma_z = 0, \tag{4.71}$$

$$\sigma_y\sigma_z + \sigma_z\sigma_y = 0, \tag{4.72}$$

$$\operatorname{tr}\sigma_x = \operatorname{tr}\sigma_y = \operatorname{tr}\sigma_z = 0, \tag{4.73}$$

$$\det\sigma_x = \det\sigma_y = \det\sigma_z = -1. \tag{4.74}$$

These properties can be easily demonstrated by using the definition of Pauli matrices, or their representation in the basis $\{|-\rangle, |+\rangle\}$. Let us calculate, for example,

$\sigma_x \sigma_y \sigma_z$:

$$\sigma_x \sigma_y \sigma_z = \begin{pmatrix} 0 & 1 \\ 1 & 0 \end{pmatrix} \begin{pmatrix} 0 & -i \\ i & 0 \end{pmatrix} \begin{pmatrix} 1 & 0 \\ 0 & -1 \end{pmatrix}$$

$$= \begin{pmatrix} 0 & 1 \\ 1 & 0 \end{pmatrix} \begin{pmatrix} 0 & i \\ i & 0 \end{pmatrix}$$

$$= \begin{pmatrix} i & 0 \\ 0 & i \end{pmatrix} = i.$$

The same result can be obtained observing that, from $[\sigma_x, \sigma_y] = \sigma_x \sigma_y - \sigma_y \sigma_x = 2i\sigma_z$ and $\{\sigma_x, \sigma_y\} = \sigma_x \sigma_y + \sigma_y \sigma_x = 0$, we have

$$\sigma_x \sigma_y = i\sigma_z$$

and, therefore,

$$\sigma_x \sigma_y \sigma_z = i\sigma_z^2 = i.$$

We leave proving the other properties of the Pauli matrices as an exercise for our readers.

5 Phase shifts and atomic potential energy

The calculation of the differential elastic scattering cross-section requires knowledge of the phase shifts. In this chapter we will provide a method for calculating the phase shifts using the Numerov algorithm, a very general numerical procedure that can be used, in particular, to solve the radial equation [15]. The calculation of the phase shifts requires knowledge of the electron density around the nucleus and of the atomic potential energy. We will see that atomic electrons have the effect of screening the potential of the bare nucleus. We will then describe two models that approximate the screened atomic potential energy: The Thomas–Fermi statistical approach and the Hartree–Fock self-consistent method [26].

5.1 An important application of the Numerov algorithm: the eigenvalue problem

Let us now consider an important application of the Numerov algorithm. Equation (3.66), i. e., the radial equation:

$$\left[\frac{d^2}{dr^2} - \frac{l(l+1)}{r^2} - U(r) + k^2 \right] F_l(r) = 0 \,,$$

has the form of Eq. (1.42). In fact, defining

$$q^2(r) = -\frac{l(l+1)}{r^2} + k^2 - U(r) = \frac{2m}{\hbar^2} \left[-\frac{l(l+1)\hbar^2}{2mr^2} + E - V(r) \right], \qquad (5.1)$$

we see that Eq. (3.66) becomes

$$\frac{d^2 F_l(r)}{dr^2} + q^2(r) F_l(r) = 0 \,. \qquad (5.2)$$

This equation has the form of Eq. (1.42) with $x = r$, $y = F_l(r)$, $f(r) = q(r)$, and $g(r) = 0$. As a consequence, we can apply to this equation the Numerov algorithm, so that

$$\left(1 + \frac{h^2}{12} q_{n+1}^2 \right)(F_l)_{n+1} - 2\left(1 - \frac{5h^2}{12} q_n^2 \right)(F_l)_n + \left(1 + \frac{h^2}{12} q_{n-1}^2 \right)(F_l)_{n-1}$$

$$= \mathcal{O}(h^6) \approx 0 \,. \qquad (5.3)$$

5.2 Phase-shift calculation

Let us indicate with r_{max} the distance from the center of the atom where the atomic potential becomes so small that it can be considered negligible. A reasonable value is

https://doi.org/10.1515/9783110675375-005

$r_{max} \approx 2-3\,\text{Å}$. Since

$$krj_l(kr) \underset{r\to\infty}{\sim} \sin\left(kr - \frac{\pi l}{2}\right), \tag{5.4}$$

$$krn_l(kr) \underset{r\to\infty}{\sim} -\cos\left(kr - \frac{\pi l}{2}\right), \tag{5.5}$$

and

$$F_l(r) \underset{r\to\infty}{\sim} \sin\left(kr - \frac{\pi l}{2} + \eta_l\right)$$
$$= \sin\left(kr - \frac{\pi l}{2}\right)\cos\eta_l + \cos\left(kr - \frac{\pi l}{2}\right)\sin\eta_l, \tag{5.6}$$

if we consider a distance r_1 from the center of the atom higher than r_{max}, then

$$F_l(r_1) \sim kr_1\left[\cos\eta_l\, j_l(kr_1) - \sin\eta_l\, n_l(kr_1)\right]. \tag{5.7}$$

Similarly, if $r_2 > r_1$, thus

$$F_l(r_2) \sim kr_2\left[\cos\eta_l\, j_l(kr_2) - \sin\eta_l\, n_l(kr_2)\right]. \tag{5.8}$$

Let us now define the quantity:

$$G = G(r_1, r_2) = \frac{r_1 F_l(r_2)}{r_2 F_l(r_1)} = \frac{\cos\eta_l\, j_l(kr_2) - \sin\eta_l\, n_l(kr_2)}{\cos\eta_l\, j_l(kr_1) - \sin\eta_l\, n_l(kr_1)}. \tag{5.9}$$

Simple algebraical manipulations allow us to conclude that

$$\tan\eta_l = \frac{G(r_1, r_2)j_l(kr_1) - j_l(kr_2)}{G(r_1, r_2)n_l(kr_1) - n_l(kr_2)}. \tag{5.10}$$

Equation (5.3) is a recursion formula for numerically integrating the radial equation (3.66) either forward or backward in r. We can use it to integrate forward in r to the radius r_1 to obtain $F_l(r_1)$ and to the radius r_2 to obtain $F_l(r_2)$. After that, we can easily calculate $G(r_1, r_2)$ and, using Eq. (5.10), the phase shifts η_l. The two integrations of Eq. (3.66) require knowledge of the atomic potential energy as a function of r, i. e., the distance from the center of the atomic nucleus. The calculation of the atomic potential energy as a function of r will be discussed in the next section.

5.3 Atomic electron density and atomic potential energy

5.3.1 Thomas–Fermi model

The simplest theory to approximate the atomic electron density and the atomic potential energy is the Thomas–Fermi statistical model. This model considers atomic electrons as a fully degenerate gas surrounding the nucleus, assumes radial symmetry, and uses Poisson's equation to obtain a differential equation known as the Thomas–Fermi equation. Once solved, it provides a dimensionless function that approximates the screening of the nucleus by the orbital electrons. This is the so-called screening function. The Coulomb potential energy of the bare nucleus multiplied by the screening function provides the atomic potential energy. Since the Thomas–Fermi model is statistical, the bigger the number of atomic electrons, the better the approximation.

A Fermi gas is a gas of non-interacting free electrons confined in a specified region of space. To describe the motion of the particles of a Fermi gas, we have to solve the Schrödinger equation:

$$-\frac{\hbar^2}{2m}\left(\frac{\partial^2}{\partial x^2} + \frac{\partial^2}{\partial y^2} + \frac{\partial^2}{\partial z^2}\right)\psi = E\psi, \tag{5.11}$$

where E is the electron energy and the wave function ψ depends on the coordinates x, y, and z, so that $\psi = 0$ at the boundaries of the region of space where the electron gas is confined. It is now convenient to factorize the wave function so that

$$\psi(x,y,z) = \psi_1(x)\,\psi_2(y)\,\psi_3(y). \tag{5.12}$$

As a consequence,

$$-\frac{\hbar^2}{2m}\frac{\partial^2}{\partial x^2}\psi_1(x) = E_1\,\psi_1(x), \tag{5.13}$$

$$-\frac{\hbar^2}{2m}\frac{\partial^2}{\partial y^2}\psi_2(y) = E_2\,\psi_2(y), \tag{5.14}$$

$$-\frac{\hbar^2}{2m}\frac{\partial^2}{\partial z^2}\psi_3(z) = E_3\,\psi_3(z), \tag{5.15}$$

and

$$E = E_1 + E_2 + E_3. \tag{5.16}$$

Assuming that

$$\psi_1(x) = 0 \quad \text{for } x \leq 0 \text{ and } x \geq l \tag{5.17}$$

and

$$\int_0^l |\psi_1(x)|^2 \, dx = 1, \tag{5.18}$$

we obtain, for

$$n_1 = 1, 2, \dots, \infty, \tag{5.19}$$

the eigenfunctions (normalized to the unity):

$$\psi_1(x) = \sqrt{\frac{2}{l}} \sin \frac{n_1 \pi x}{l}, \tag{5.20}$$

and the eigenvalues:

$$E_1 = \frac{1}{2m} \left(\frac{\pi \hbar}{l} \right)^2 n_1^2. \tag{5.21}$$

In fact, if

$$\psi_1(x) = A \sin k_1 x, \tag{5.22}$$

where A is a constant to be determined, then

$$-\frac{\hbar^2}{2m} (-k_1^2 A \sin k_1 x) = E_1 A \sin k_1 x, \tag{5.23}$$

or

$$E_1 = \frac{\hbar^2 k_1^2}{2m}. \tag{5.24}$$

Since $A \sin k_1 l = 0$,

$$k_1 = \frac{n_1 \pi}{l}. \tag{5.25}$$

From

$$1 = A^2 \int_0^l \sin^2 k_1 x \, dx = \frac{A^2}{k_1} \frac{1}{2} (k_1 l - \sin k_1 l \cos k_1 l) = \frac{A^2}{2} l, \tag{5.26}$$

we obtain

$$A = \sqrt{\frac{2}{l}}. \tag{5.27}$$

Once the following two new series of integers:

$$n_2 = 1, 2, \ldots, \infty \qquad (5.28)$$

and

$$n_3 = 1, 2, \ldots, \infty, \qquad (5.29)$$

have been introduced, we can conclude that, assuming that the electron gas is confined in a cubic box of side l, the wave function is given by

$$\psi(x,y,z) = \left(\frac{2}{l}\right)^{3/2} \sin\frac{n_1\pi x}{l} \sin\frac{n_2\pi y}{l} \sin\frac{n_3\pi z}{l}, \qquad (5.30)$$

with eigenvalues of energy

$$E = \frac{1}{2m}\left(\frac{\pi\hbar}{l}\right)^2 (n_1^2 + n_2^2 + n_3^2). \qquad (5.31)$$

We now wish to answer the following question: What is the number of states whose kinetic energy ranges between E and $E + dE$? To answer this question, let's introduce the radius vector $\mathbf{w}$, from the origin of a three-dimensional space, such that

$$w^2 = n_1^2 + n_2^2 + n_3^2 \qquad (5.32)$$

and, consequently,

$$E = \frac{1}{2m}\left(\frac{\pi\hbar}{l}\right)^2 w^2. \qquad (5.33)$$

To proceed, let's now introduce an approximation. We assume that the components of vector $\mathbf{w}$ are continuous rather than discrete, so

$$dE = \frac{1}{m}\left(\frac{\pi\hbar}{l}\right)^2 w\,dw. \qquad (5.34)$$

Since the components of $\mathbf{w}$ are positive, the number of states $N(E)\,dE$ is equal to 1/8 of the volume of space delimited by two spherical shells having radius, respectively, $\mathbf{w}$ and $\mathbf{w} + \mathbf{dw}$, so that

$$N(E)\,dE = \frac{1}{8}d\left(\frac{4}{3}\pi w^3\right) = \frac{\pi w^2}{2}dw \qquad (5.35)$$

and, therefore,

$$N(E)\,dE = \frac{\Omega\, m^{3/2}}{\sqrt{2}\,\hbar^3\,\pi^2}\sqrt{E}\,dE, \qquad (5.36)$$

where Ω is the volume of the region of space where the Fermi gas is confined. Note that $\Omega = l^3$ if the region of space where electrons are confined is cubic, as in this example. Actually, the shape of the region where the non-interacting free electrons are confined is not relevant. Equation (5.36) is always valid, regardless of the shape of the confinement region.

Due to Pauli's exclusion principle, each energetic level of the Fermi gas cannot be occupied by more than two electrons, corresponding to the two possible spin orientations. The highest electron energy of the Fermi gas at $T = 0$ (T being the absolute temperature) is the so-called Fermi energy E_F, defined by

$$N_{TOT} = 2 \int_0^{E_F} N(E) \, dE, \tag{5.37}$$

where we have indicated with N_{TOT} the total number of electrons confined in the region of space with volume Ω. The electron density ρ is given by

$$\rho = \frac{N_{TOT}}{\Omega}. \tag{5.38}$$

The factor 2 in front of the integral over all the states in Eq. (5.37) is due to the fact that each energetic level, at $T = 0$, contains two electrons with opposite spins.

By using Eqs. (5.36) and (5.37), we obtain the dependence of Fermi energy E_F on electron density ρ:

$$E_F = \frac{\hbar^2}{2m} (3\pi^2 \rho)^{2/3}. \tag{5.39}$$

The Thomas–Fermi model is based on the assumption that the atomic electrons of an atom with many electrons can be approximately described as a Fermi gas at $T = 0$, confined by a potential energy $-V(r)$, in a spherically shaped region. Therefore, at distance r from the nucleus, we should have

$$-V(r) = \frac{\hbar^2}{2m} [3\pi^2 \rho(r)]^{2/3}. \tag{5.40}$$

Let us indicate the electric field with $\mathcal{E}(r)$, so that:

$$e\,\mathcal{E}(r) = \frac{dV(r)}{dr}, \tag{5.41}$$

where e is the elementary charge. According to Gauss's law:

$$\int_S \mathcal{E}(r) \cdot d\mathbf{s} = 4\pi\,Q, \tag{5.42}$$

where Q is the total charge contained in the closed surface S. In our case, the surface is a sphere whose radius is r and the electric field, for any given radius r, is constant and normal to the surface, so

$$\int_S \mathcal{E}(r) \cdot d\mathbf{s} = 4\pi r^2 \, \mathcal{E}(r).$$ (5.43)

Indicating with Z the atomic number, the total charge is given by the nuclear charge Ze plus the electronic charge $-\int_0^r 4\pi e\rho(r')r'^2 dr'$. As a consequence,

$$r^2 \mathcal{E}(r) = Ze - 4\pi e \int_0^r \rho(r')r'^2 \, dr',$$ (5.44)

and, therefore,

$$\frac{1}{r^2}\frac{d}{dr}\left[r^2\frac{dV(r)}{dr}\right] = -4\pi e^2 \rho(r).$$ (5.45)

Taking into account Eq. (5.40), we can easily obtain the dependence on r of the electron density $\rho(r)$:

$$\rho(r) = \left(\sqrt{\frac{2m}{\hbar^2}}\right)^3 \frac{1}{3\pi^2} [-V(r)]^{3/2}.$$ (5.46)

As a consequence, the atomic potential energy of a Thomas–Fermi atom must satisfy the following differential equation:

$$\frac{1}{r^2}\frac{d}{dr}\left\{r^2\frac{d}{dr}[-V(r)]\right\} = \frac{8\sqrt{2}\,e^2\sqrt{m^3}}{3}\frac{}{\pi\hbar^3}\sqrt{[-V(r)]^3}.$$ (5.47)

At great distances from the center of the nucleus, the potential energy must approach zero:

$$\lim_{r\to\infty} r\,[-V(r)] = 0,$$ (5.48)

while, since when r is small $V(r)$ should approach the potential energy of the nucleus, we have

$$\lim_{r\to 0} r\,[-V(r)] = Ze^2.$$ (5.49)

The previous two equations represent boundary conditions that must be satisfied when solving the differential Eq. (5.47). Let us now introduce the Bohr radius a_0:

$$a_0 = \frac{\hbar^2}{me^2},$$ (5.50)

and the Thomas–Fermi radius a_{TF}:

$$a_{TF} = \frac{1}{2}\left(\frac{3\pi}{4}\right)^{2/3}\frac{a_0}{Z^{1/3}}.$$

(5.51)

Once the auxiliary variable x:

$$x = \frac{r}{a_{TF}}$$

(5.52)

and the auxiliary function ξ:

$$\xi = \frac{r[-V(r)]}{Ze^2}$$

(5.53)

have been defined, from Eq. (5.47) we obtain the differential equation:

$$\frac{d^2\xi}{dx^2} = \frac{\xi^{3/2}}{x^{1/2}}$$

(5.54)

that is subject to the boundary conditions [see Eqs. (5.48) and (5.49)]:

$$\lim_{x\to0}\xi(x) = 1,$$

(5.55)

$$\lim_{x\to\infty}\xi(x) = 0.$$

(5.56)

Indeed

$$\frac{1}{r^2}\frac{d}{dr}\left\{r^2\frac{d}{dr}[-V(r)]\right\} = \frac{1}{r^2}\frac{d}{dr}\left\{r^2\frac{d}{dr}\left[\frac{Ze^2}{r}\xi\right]\right\}$$

$$= \frac{1}{r^2}\frac{d}{dr}\left[r^2Ze^2\left(-\frac{1}{r^2}\right)\xi + r^2\frac{Ze^2}{r}\frac{d\xi}{dr}\right]$$

$$= \frac{1}{r^2}\frac{d}{dr}\left[-Ze^2\xi + rZe^2\frac{d\xi}{dr}\right] = \frac{Ze^2}{r}\frac{d^2\xi}{dr^2},$$

so that

$$\frac{Ze^2}{r}\frac{d^2\xi}{dr^2} = \frac{8\sqrt{2}e^2m^{3/2}}{3\pi\hbar^3}\left(\frac{Ze^2}{r}\xi\right)^{3/2},$$

or

$$\frac{d^2\xi}{dr^2} = \frac{8\sqrt{2}e^2m^{3/2}}{3\pi\hbar^3}\left(\frac{Ze^2}{r}\right)^{1/2}\xi^{3/2}.$$

Now, since

$$r = a_{TF}x,$$

we can also express our equation as a function of x as

$$\frac{1}{a_{TF}^2} \frac{d^2\xi}{dx^2} = \frac{8\sqrt{2}e^2 m^{3/2}}{3\pi\hbar^3} (Ze^2)^{1/2} \frac{\xi^{3/2}}{a_{TF}^{1/2} x^{1/2}}.$$

From the definition of the Thomas–Fermi radius a_{TF}, Eq. (5.51), we can easily see that

$$\frac{8\sqrt{2}e^3 m^{3/2} Z^{1/2}}{3\pi\hbar^3} a_{TF}^{3/2} = 1,$$

and, as a consequence,

$$\frac{d^2\xi}{dx^2} = \left[\frac{8\sqrt{2}e^3 m^{3/2} Z^{1/2}}{3\pi\hbar^3} a_{TF}^{3/2}\right] \frac{\xi^{3/2}}{x^{1/2}} = \frac{\xi^{3/2}}{x^{1/2}}.$$

Function ξ is the so-called screening function. In the Thomas–Fermi model, it describes the screening of the atomic potential due to the cloud of atomic electrons. The solution to Eq. (5.54), subject to the boundary conditions given by Eqs. (5.55) and (5.56), enables us to obtain the Thomas–Fermi atomic potential energy by multiplying the potential energy of the bare nucleus by the screening function (5.53):

$$-V(r) = \frac{Ze^2}{r}\xi. \tag{5.57}$$

5.3.2 Hartree and Hartree–Fock approximations

In the previous section we discussed the statistical Thomas–Fermi atomic model. Being statistical, it is not surprising that the Thomas–Fermi method is inaccurate for $Z < 10$. Also, the outer layers of all atoms are not well described by the Thomas–Fermi approach.

In this section we will discuss more accurate methods for calculating the atomic potential energy as a function of the distance from the center of the nucleus. They are the Hartree and the Hartree–Fock theories, in which the Schrödinger equation is solved by successive approximations subject to the requirement of self-consistency (self-consistent field).

The Hartree approximation
Let us consider an atom with N electrons and a nucleus with charge Ze. Let $\psi(1,\ldots,N)$ be the wave function describing the system, and let us write the Schrödinger equation as

$$H\psi(1,\ldots,N) = E\psi(1,\ldots,N), \tag{5.58}$$

where E is the atom total energy,

$$H = \sum_{i=1}^{N} \left[\frac{\mathbf{p}_i^2}{2m} + U(\mathbf{r}_i) \right] + \sum_{i>j} \frac{e^2}{r_{ij}} \tag{5.59}$$

is the Hamiltonian,

$$U(\mathbf{r}_i) = -\frac{Ze^2}{r_i} \tag{5.60}$$

is the electron–nucleus potential energy, e^2/r_{ij} are the potential energies due to the electron–electron interactions,

$$r_i = |\mathbf{r}_i| \tag{5.61}$$

are the electron–nucleus distances, and

$$r_{ij} = |\mathbf{r}_i - \mathbf{r}_j| \tag{5.62}$$

are the electron–electron distances. The Hartree approximation is based on the factorization of the wave function:

$$\psi(1,\dots,N) = \psi_1(1)\,\psi_2(2) \cdots \psi_N(N), \tag{5.63}$$

where

$$\psi_i(i) = \psi(\mathbf{r}_i)\chi_i(m_{s_i}) \tag{5.64}$$

are the products of spatial wave functions by spin states. Furthermore, we assume that functions $\psi_i(\mathbf{r}_i)$ satisfy the normalization conditions:

$$\langle\psi_i|\psi_i\rangle = \int |\psi_i(\mathbf{r}_i)|^2 d^3r_i = 1. \tag{5.65}$$

The expectation value of the Hamiltonian H is given by

$$\langle H \rangle = \sum_{i=1}^{N} \left\{ \int d^3r_i\, \psi_i^*(\mathbf{r}_i) \left[-\frac{\hbar^2}{2m}\nabla_i^2 - \frac{Ze^2}{r_i} \right] \psi_i(\mathbf{r}_i) \right\}$$
$$+ \sum_{i>j} \int d^3r_i \int d^3r_j\, \psi_j^*(\mathbf{r}_j)\psi_i^*(\mathbf{r}_i) \frac{e^2}{r_{ij}} \psi_i(\mathbf{r}_i)\psi_j(\mathbf{r}_j). \tag{5.66}$$

Since we aim at minimizing $\langle H \rangle$ with the constraint of normalization expressed by Eq. (5.65), we have to calculate the functional derivatives of the functional:

$$\mathcal{F} = \langle H \rangle - \sum_{i=1}^{N} \varepsilon_i \left(\int |\psi_i(\mathbf{r}_i)|^2 d^3 r_i - 1 \right), \qquad (5.67)$$

where ε_i are N Lagrange multipliers—one per atomic electron—to be determined.

The expectation value of the Hamiltonian has to be regarded as a functional of the single-particle wave functions and it is required to be stationary with respect to all the possible variations of the wave functions describing the single particles, subject to the constraint of norm-conservation.

Then, the functional derivative of $\mathcal{F}$ with respect to $\psi_i^*(\mathbf{r}_i)$ has to be zero. Therefore, we obtain the N Hartree equations:

$$\left[-\frac{\hbar^2}{2m} \nabla_i^2 - \frac{Ze^2}{r_i} + V_i(\mathbf{r}_i) \right] \psi_i(\mathbf{r}_i) = \varepsilon_i \psi_i(\mathbf{r}_i), \qquad (5.68)$$

where the electrostatic potential energy $V_i(\mathbf{r}_i)$ due to the electron cloud is given by

$$V_i(\mathbf{r}_i) = e^2 \sum_{i \neq j} \int d^3 r_j \frac{1}{r_{ij}} |\psi_j(\mathbf{r}_j)|^2. \qquad (5.69)$$

The behavior of ith wave function ψ_i is determined by V_i, a function that depends on all the other atomic electron wave functions. Basically, the first set of trial functions V_i is obtained by using Eq. (5.69) with a set of reasonable trial functions ψ_j. Once the V_i functions have been obtained, they are used to solve Eq. (5.68).

In such a way we obtain a new set of ψ_i functions. The new set of functions ψ_i is then used in Eq. (5.69) to obtain a better approximation of the set of functions V_i, and so on.

This procedure can be interrupted when there are no further significant changes in both the set of functions $V_i(\mathbf{r}_i)$ and $\psi_i(\mathbf{r}_i)$.

Once the self-consistent calculation has been concluded, the final $V_i(\mathbf{r}_i)$ and $\psi_i(\mathbf{r}_i)$ functions are used to obtain the values of the Lagrange multipliers ε_i (i. e., the ionization energies) and the total energy (i. e., the expectation value of H obtained using the numerically obtained wave function ψ). It is given by

$$\langle \psi | H | \psi \rangle = E = \sum_{i=1}^{N} \varepsilon_i - e^2 \sum_{i < j} \int \frac{1}{r_{ij}} |\psi_i(\mathbf{r}_i)|^2 |\psi_j(\mathbf{r}_j)|^2 d^3 r_i \, d^3 r_j. \qquad (5.70)$$

The Hartree–Fock approximation
In the Hartree approximation, the wave function is the simple product of all the single-particle wave functions. The Hartree–Fock approximation also takes into account that the total wave function must obey the Pauli exclusion principle, which states that two

identical fermions cannot occupy the same quantum state. So, in the Hartree–Fock approximation, the trial function including the spin is a Slater determinant:

$$\psi(1,\ldots,N) = (N!)^{-\frac{1}{2}} \begin{vmatrix} \psi_1(1) & \cdots & \psi_1(N) \\ & \cdot & \\ \cdot & \cdot & \cdot \\ & \cdot & \\ \psi_N(1) & \cdots & \psi_N(N) \end{vmatrix}, \tag{5.71}$$

where

$$\psi_i(j) = \psi_i(\mathbf{r}_j)\chi_i(m_{s_j}), \tag{5.72}$$

and $\chi_i(m_{s_j})$ are the spin states.

A function with the form of a Slater determinant obeys the Pauli exclusion principle: Indeed, it has the required properties of antisymmetry with respect to the interchange of any pair of electrons. Furthermore, ψ is correctly normalized if the single-particle functions are orthonormal, i. e., if

$$\langle \psi_i | \psi_j \rangle = \delta_{ij}. \tag{5.73}$$

If we use the Hartree method, but replace the simple factorization utilized by the Hartree approximation with the Slater determinant, we obtain the Hartree–Fock equations:

$$\left[-\frac{\hbar^2}{2m}\nabla_i^2 - \frac{Ze^2}{r_i} \right]\psi_i(\mathbf{r}_i)$$

$$+ e^2 \int d^3r_j \frac{1}{r_{ij}} \sum_{j=1}^{N} \psi_j^*(\mathbf{r}_j)[\psi_j(\mathbf{r}_j)\psi_i(\mathbf{r}_i) - \psi_j(\mathbf{r}_i)\psi_i(\mathbf{r}_j)\delta_{m_{s_i}m_{s_j}}]$$

$$= \varepsilon_i\psi_i(\mathbf{r}_i). \tag{5.74}$$

The Hartree–Fock set of equations can be solved by iteration, similarly to what occurs in the Hartree approach. Note that the Hartree–Fock set of equations contain a non-local term (where the argument of ψ_i is $\mathbf{r}_j$ and the argument of ψ_j is $\mathbf{r}_i$) known as the exchange term and different from zero only when $m_{s_i} = m_{s_j}$.

Part III: **Quantum-relativistic equations and scattering**

6 The Klein–Gordon and the Dirac relativistic equations

The first attempts of generalization of the Schrödinger equation to obtain a quantum relativistic equation were performed by Schrödinger himself and by Oskar Klein and Walter Gordon in 1926–1927. The first equation proposed in those years is the so-called Klein–Gordon equation. We will see that the Klein–Gordon equation predicts negative energy states and allows that probability densities can be also negative. Thus, it was soon abandoned. The Klein–Gordon equation was subsequently reinterpreted as a fundamental equation in quantum field theory. In particular, in 1934, Wolfgang Pauli and Victor Weisskopf demonstrated that it correctly describes particles with null spin, such as π mesons. The temporary dismissal of the Klein–Gordon equation prompted renewed effort and further research. These new studies made possible the discovery, in 1928, of another quantum-relativistic equation by Paul Adrien Maurice Dirac. The Dirac equation, even if does not eliminate the negative energy states, correctly describes particles with spin 1/2, such as electrons and positrons. The existence of two quantum-relativistic equations with similar difficulties suggested the idea that the equations were correct and that they required a reinterpretation. The final result was that it was necessary to modify the fundamental concepts about the properties of the particles, of the fields, and of the radiation–matter interaction [30, 33]. This chapter is devoted to the deduction of the Klein–Gordon and the Dirac equations. We will also present their main properties and study their characteristics [9, 18, 21, 27, 28].

6.1 The natural system of units

From now on, we will use the system of units defined by $\hbar = 1$ and $c = 1$, where $\hbar = h/2\pi$ is the reduced Planck constant and c is the speed of light in a vacuum. This system of units is known as the *natural system*. In the natural system only one unit, among energy, time, and length, is independent. Time has the dimension of a length as a consequence of the fact that c is the ratio between length and time. Energy has the dimension of an inverse length as a consequence of the fact that $\hbar$ is the product of energy by time. As in atomic physics, the usual energy unit is eV. So, length is measured in eV^{-1}. Since time has the dimension of a length, it is also measured in eV^{-1}. Mass, due to the Einstein law of equivalence between mass and energy, is measured in eV. Like $\hbar$ and c, also the elementary electric charge e is a dimensionless quantity, e. g., $e^2 = e^2/(\hbar c) \approx 1/137$. To switch to ordinary units, we just need to include in the equations the required powers of $\hbar$ and c. For example, the relativistic equation relating energy and momentum is expressed, in natural units, as $E^2 = p^2 + m^2$, and, in ordinary units, as $E^2 = p^2 c^2 + m^2 c^4$.

https://doi.org/10.1515/9783110675375-006

6.2 The Lorentz transformation

Here, we would like to remind our readers about the origin of the so-called *Lorentz transformation*, which relates the rest frame coordinates t, x, y, z to the coordinates t', x', y', z' of a frame moving with constant velocity v in the direction of the x axis [9, 16, 21, 28, 29]. Let us first observe that space-time is everywhere the same, so there are no special events in space-time. As a consequence, the transformation has to be linear. Since x' has to be zero whenever $x = vt$, we can write

$$x' = (x - vt)f, \tag{6.1}$$

where f is a function of the constant velocity v of the moving frame. Actually, since there are not preferred directions in space, f has to be a function of v^2:

$$f = f(v^2). \tag{6.2}$$

Similarly, since t' has to be zero whenever $t = vx$, we can write [29]

$$t' = (t - vx)g. \tag{6.3}$$

Let us now consider a light ray. The speed of light is $c = 1$, so, in the rest frame, the path of the light ray is described by the equation $x = t$. Since the speed of light does not depend on the reference frame, the path of the light ray is described, in the moving frame, by the equation $x' = t'$. In other words, if $x = t$, then $x' = t'$. Let us set $x = t$ in Eqs. (6.1) and (6.3):

$$x' = (t - vt)f(v^2),$$
$$t' = (t - vt)g(v^2).$$

From $x' = t'$ we then obtain:

$$g(v^2) = f(v^2). \tag{6.4}$$

Of course, these arguments can be inverted. If the "moving frame" is considered at rest, the "rest frame" is moving with velocity $-v$. So, in conclusion, we can write [29]

$$x' = (x - vt)f(v^2), \tag{6.5}$$
$$t' = (t - vx)f(v^2), \tag{6.6}$$
$$x = (x' + vt')f(v^2), \tag{6.7}$$
$$t = (t' + vx')f(v^2). \tag{6.8}$$

From these last equations we obtain, in particular,

$$t = [(t - vx) + v(x - vt)]f^2(v^2) = tf^2(v^2)(1 - v^2),$$

and hence

$$f^2(v^2) = \frac{1}{1 - v^2}.$$ (6.9)

So, in conclusion,

$$t' = \frac{t - vx}{\sqrt{1 - v^2}},$$ (6.10)

$$x' = \frac{x - vt}{\sqrt{1 - v^2}}.$$ (6.11)

As for the coordinates y and z, they are passive, being the relative motion along the x axis. For the reader's convenience, we write the Lorentz transformation for a relative motion along the x axis in conventional units:

$$t' = \frac{t - vx/c^2}{\sqrt{1 - v^2/c^2}},$$ (6.12)

$$x' = \frac{x - vt}{\sqrt{1 - v^2/c^2}},$$ (6.13)

$$y' = y,$$ (6.14)

$$z' = z.$$ (6.15)

From the Lorentz transformation the invariance of the interval, also known as the *proper time*, immediately follows. Indeed,

$$t'^2 - x'^2 - y'^2 - z'^2 = \frac{(t - vx)^2 - (x - vt)^2}{1 - v^2} - y^2 - z^2$$

$$= (t^2 - x^2)\frac{1 - v^2}{1 - v^2} - y^2 - z^2 = t^2 - x^2 - y^2 - z^2.$$

So, the value of the square of the proper time:

$$\tau^2 = t^2 - x^2 - y^2 - z^2$$ (6.16)

is an invariant because it does not change under any Lorentz transformation.

6.3 Four-vectors and tensors

In space-time, an event is specified by the four coordinates x^0, x^1, x^2, x^3, where $x^0 = ct = t$ is the time coordinate and $x^1 = x, x^2 = y, x^3 = z$ are the three spatial coordinates. The components of four-vectors and tensors along the four axes t, x, y, z are specified by the use of the indices $0,1,2,3$, respectively. We will use Greek indices (such as $\mu, \nu, \rho, \ldots$) to denote the components of four-vectors of space-time and

Roman indices (such as $j,k,l,...$) to denote the components of three-vectors of the ordinary space. An event in space-time is, in particular, a four-vector and, therefore,

$$x^\mu = (x^0, x^j) = (x^0, x^1, x^2, x^3) = (t, x, y, z), \tag{6.17}$$

where $\mu = 0,1,2,3$ and $j = 1,2,3$. Let us consider a Lorentz transformation along the x axis. Once introduced, the Lorentz matrix (describing a *boost* along the x axis):

$$\Lambda^\mu{}_\nu = \begin{pmatrix} \frac{1}{\sqrt{1-v^2}} & \frac{-v}{\sqrt{1-v^2}} & 0 & 0 \\ \frac{-v}{\sqrt{1-v^2}} & \frac{1}{\sqrt{1-v^2}} & 0 & 0 \\ 0 & 0 & 1 & 0 \\ 0 & 0 & 0 & 1 \end{pmatrix}, \tag{6.18}$$

or

$$\Lambda^\mu{}_\nu = \begin{pmatrix} \gamma & -\gamma v & 0 & 0 \\ -\gamma v & \gamma & 0 & 0 \\ 0 & 0 & 1 & 0 \\ 0 & 0 & 0 & 1 \end{pmatrix}, \tag{6.19}$$

where

$$\gamma = \frac{1}{\sqrt{1-v^2}}, \tag{6.20}$$

we can write

$$x'^\mu = \sum_{\nu=0}^{3} \Lambda^\mu{}_\nu x^\nu. \tag{6.21}$$

In general we have

$$dx'^\mu = \sum_{\nu=0}^{3} \frac{\partial x'^\mu}{\partial x^\nu} dx^\nu. \tag{6.22}$$

If x'^μ and x^μ are related by linear equations, as in Lorentz transformations, we can write

$$x'^\mu = \sum_{\nu=0}^{3} \frac{\partial x'^\mu}{\partial x^\nu} x^\nu, \tag{6.23}$$

so that

$$\Lambda^\mu{}_\nu = \frac{\partial x'^\mu}{\partial x^\nu}. \tag{6.24}$$

Although in this book we will consider only Lorentz transformations, we observe that the most general transformation between inertial frames of reference is the *Poincaré transformation*:

$$x'^\mu = \sum_{\nu=0}^{3} \Lambda^\mu{}_\nu x^\nu + s^\mu, \qquad (6.25)$$

where s^μ are constants. The Lorentz transformation is the Poincaré transformation for the special case $s^\mu = 0$.

A *contravariant* four-vector a^μ (indices are indicated as superscripts) transforms like x^μ, while a *covariant* four-vector a_μ (indices are indicated as subscripts) transforms like $\partial/\partial x^\mu$. The correspondence between covariant and contravariant components of a four-vector is ruled by the *metric tensor* $g_{\mu\nu}$ defined by

$$g_{\mu\nu} = \begin{pmatrix} 1 & 0 & 0 & 0 \\ 0 & -1 & 0 & 0 \\ 0 & 0 & -1 & 0 \\ 0 & 0 & 0 & -1 \end{pmatrix}. \qquad (6.26)$$

The following equation is used to obtain the covariant components of a four-vector from its contravariant components:

$$a_\mu = g_{\mu\nu} a^\nu. \qquad (6.27)$$

Please note that, in the previous equation, we have followed the so-called *Einstein convention of summing over repeated indices*. According to this convention,

$$g_{\mu\nu} a^\nu = \sum_{\nu=0}^{3} g_{\mu\nu} a^\nu. \qquad (6.28)$$

In other words, if a Greek index is repeated, once at the bottom and once at the top, then the sum on that index from 0 to 3 is implied. Please note that, to raise an index of a four-vector, we apply a similar rule, namely,

$$a^\mu = g^{\mu\nu} a_\nu. \qquad (6.29)$$

Using the Einstein convention of summing over repeated indices, the Lorentz transformation, Eq. (6.21), becomes

$$x'^\mu = \Lambda^\mu{}_\nu x^\nu. \qquad (6.30)$$

Since a contravariant four-vector a^μ transforms like x^μ, we have

$$a'^\mu = \Lambda^\mu{}_\nu a^\nu. \qquad (6.31)$$

The components of a tensor transform like the products of the components of vectors. We use the expression "tensor $t^{\mu\nu}$" instead of the stricter expression "tensor whose contravariant components are $t^{\mu\nu}$". Please note that the rules just introduced to lower and raise indices of four-vectors also apply to tensors. For example,

$$t_\mu{}^\nu = g_{\mu\rho}\, t^{\rho\nu}. \tag{6.32}$$

The *contraction* operation on a tensor consists of setting a contravariant index equal to a covariant index, implying Einstein's convention on repeated indices. If contraction is done in combination with multiplication, it is called *internal multiplication*. The contraction:

$$b_\mu = t_{\mu\nu}\, a^\nu \tag{6.33}$$

is an example of internal multiplication. If an equation has a tensor nature, it takes the same form in any inertial frame of reference. Tensor equations are invariant in form. The equations expressed in this form are called *manifestly covariant*.

The well-known Kronecker symbol $\delta_\mu{}^\nu$, equal to 1 if $\mu = \nu$ and to 0 otherwise, is obtained by lowering an index of the metric tensor, according to the rule:

$$\delta_\mu{}^\nu \equiv g_\mu{}^\nu = g_{\mu\rho}\, g^{\rho\nu} = \begin{cases} 1 & \text{if } \mu = \nu, \\ 0 & \text{if } \mu \neq \nu. \end{cases} \tag{6.34}$$

By definition,

$$\begin{aligned}
a_0 &= g_{00}\, a^0 = a^0, \\
a_1 &= g_{11}\, a^1 = -a^1, \\
a_2 &= g_{22}\, a^2 = -a^2, \\
a_3 &= g_{33}\, a^3 = -a^3.
\end{aligned} \tag{6.35}$$

The scalar product of two four-vectors a^μ and b^ν is given by

$$g_{\mu\nu} a^\mu b^\nu = a^\mu b_\mu = a_\mu b^\mu = a^0 b^0 - \mathbf{a}\cdot\mathbf{b}. \tag{6.36}$$

Therefore, the norm of a four-vector a^μ is given by

$$a^\mu a_\mu = (a^0)^2 - \mathbf{a}^2. \tag{6.37}$$

Let us now consider the matrix obtained by lowering the first index and raising the
second index of the Lorentz matrix $\Lambda^\mu{}_\nu$:[1]

$$\Lambda_\mu{}^\nu = \begin{pmatrix} \gamma & \gamma v & 0 & 0 \\ \gamma v & \gamma & 0 & 0 \\ 0 & 0 & 1 & 0 \\ 0 & 0 & 0 & 1 \end{pmatrix}. \tag{6.41}$$

Using Eqs. (6.19) and (6.41), we immediately see that $\Lambda_\mu{}^\nu$ is the inverse of $\Lambda^\mu{}_\nu$, i. e.:[2]

$$\Lambda^\mu{}_\nu \, \Lambda_\mu{}^\rho = \delta_\nu{}^\rho . \tag{6.42}$$

As a consequence, if a^μ and b^μ are two four-vectors, then their scalar product is
Lorentz-invariant. In fact,

$$a'^\mu \, b'_\mu = \Lambda^\mu{}_\nu \, \Lambda_\mu{}^\rho \, a^\nu \, b_\rho = \delta_\nu{}^\rho \, a^\nu \, b_\rho = a^\rho \, b_\rho = a^\mu \, b_\mu . \tag{6.43}$$

Please note that $d\tau^2$, in particular, is the Lorentz-invariant norm of the four-vector dx^μ:

$$d\tau^2 = g_{\mu\nu} dx^\mu \, dx^\nu = dx^\mu \, dx_\mu = dt^2 - dx^2 - dy^2 - dz^2 . \tag{6.44}$$

An important example of a four-vector is the partial-differentiation operator $\partial/\partial x^\mu$. It
is a covariant four-vector denoted by the symbol ∂_μ:

$$\partial_\mu = \frac{\partial}{\partial x^\mu} . \tag{6.45}$$

---

1 Please note that, even if the Lorentz matrix is not a tensor, the rules about lowering and raising
indices are the same as those that apply to tensors. For example:

$$\Lambda_{\nu\rho} = g_{\mu\nu} \Lambda^\mu{}_\rho = \begin{pmatrix} \gamma & -\gamma v & 0 & 0 \\ \gamma v & -\gamma & 0 & 0 \\ 0 & 0 & -1 & 0 \\ 0 & 0 & 0 & -1 \end{pmatrix}. \tag{6.38}$$

Using Eqs. (6.19) and (6.38), it is then easy to verify that

$$\Lambda^\mu{}_\nu \Lambda_{\mu\rho} = g_{\nu\rho} \tag{6.39}$$

and

$$\Lambda^\mu{}_\nu \Lambda_\mu{}^\rho = g_\nu{}^\rho = \delta_\nu{}^\rho . \tag{6.40}$$

2 This also follows from the fact that the sign of v is reversed. The inverse transformation is

$$x^\mu = x'^\nu \Lambda_\nu{}^\mu .$$

We hence have

$$\partial_\mu = \left(\frac{\partial}{\partial t}, \nabla\right) = \left(\frac{\partial}{\partial t}, \frac{\partial}{\partial x}, \frac{\partial}{\partial y}, \frac{\partial}{\partial z}\right). \tag{6.46}$$

The d'Alembert operator $\square$ is the norm of ∂_μ:

$$\square = \partial_\mu \partial^\mu = \frac{\partial^2}{\partial t^2} - \nabla^2. \tag{6.47}$$

Another important examples of four-vectors is the so-called *four-potential*:

$$A^\mu = (\varphi, \mathbf{A}), \tag{6.48}$$

where φ is the scalar potential and $\mathbf{A}$ is the vector potential. A very important space-time tensor is the antisymmetric *electromagnetic tensor* $F_{\mu\nu}$ given by

$$F_{\mu\nu} = \frac{\partial A_\nu}{\partial x^\mu} - \frac{\partial A_\mu}{\partial x^\nu}. \tag{6.49}$$

It can also be expressed as

$$F_{\mu\nu} = \begin{pmatrix} 0 & \mathcal{E}_x & \mathcal{E}_y & \mathcal{E}_z \\ -\mathcal{E}_x & 0 & -\mathcal{H}_z & \mathcal{H}_y \\ -\mathcal{E}_y & \mathcal{H}_z & 0 & -\mathcal{H}_x \\ -\mathcal{E}_z & -\mathcal{H}_y & \mathcal{H}_x & 0 \end{pmatrix}, \tag{6.50}$$

where

$$\mathcal{E} = -\nabla\varphi - \frac{\partial \mathbf{A}}{\partial t} \tag{6.51}$$

is the electric field and

$$\mathcal{H} = \nabla \times \mathbf{A} \tag{6.52}$$

is the magnetic field.

6.4 The Hamiltonian of a charged particle in an electromagnetic field

Let us now consider a particle of rest mass m and electric charge e. The relativistic mass M is given by

$$M = \frac{m}{\sqrt{1 - v^2}}, \tag{6.53}$$

and the mechanical momentum $\boldsymbol{\pi}$ is

$$\boldsymbol{\pi} = M\mathbf{v} = \frac{m\mathbf{v}}{\sqrt{1 - v^2}} \,. \tag{6.54}$$

M and $\boldsymbol{\pi}$ form the four-vector π^{μ}:

$$\pi^{\mu} = (M, \boldsymbol{\pi}) \tag{6.55}$$

whose norm is the square of the rest mass m of the particle:

$$M^2 - \pi^2 = m^2 \,. \tag{6.56}$$

If the charged particle is in an electromagnetic field, then the energy E and the momentum $\mathbf{p}$ are given by, respectively,

$$E = M + e\varphi \,, \tag{6.57}$$
$$\mathbf{p} = \boldsymbol{\pi} + e\mathbf{A} \,. \tag{6.58}$$

Energy and momentum form the four-vector p^{μ} given by

$$p^{\mu} = \pi^{\mu} + e\,A^{\mu} \,. \tag{6.59}$$

From Eqs. (6.56), (6.57), and (6.58), we have

$$(E - e\varphi)^2 = (\mathbf{p} - e\mathbf{A})^2 + m^2 \,, \tag{6.60}$$

so the Hamiltonian of a particle of rest mass m and electric charge e in an electromagnetic field described by the four-potential $A^{\mu} = (\varphi, \mathbf{A})$ is given by

$$H = e\varphi + \sqrt{(\mathbf{p} - e\mathbf{A})^2 + m^2} \,. \tag{6.61}$$

6.5 Klein–Gordon equation

The Schrödinger equation expressed in natural units is given by

$$H\psi = i\frac{\partial\psi}{\partial t} \,. \tag{6.62}$$

Using the relativistic equation relating energy and momentum of a free particle:

$$E^2 = p^2 + m^2 \,, \tag{6.63}$$

and the correspondence rule

$$p^2 \rightarrow -\nabla^2 \,, \tag{6.64}$$

where

$$\nabla^2 = \frac{\partial^2}{\partial x^2} + \frac{\partial^2}{\partial y^2} + \frac{\partial^2}{\partial z^2}, \tag{6.65}$$

the Schrödinger equation assumes the form:

$$\sqrt{m^2 - \nabla^2}\, \psi = i\frac{\partial \psi}{\partial t}. \tag{6.66}$$

This equation presents an asymmetry between the spatial coordinates and the time coordinate. Furthermore, the presence of the square root operator implies a nonlocal theory. For these reasons, Oskar Klein and Walter Gordon, in 1926, proposed considering the square of the Hamiltonian to obtain

$$\left(\frac{\partial^2}{\partial t^2} - \nabla^2 + m^2 \right) \psi = 0, \tag{6.67}$$

or

$$\left(\frac{\partial^2}{\partial t^2} - \frac{\partial^2}{\partial x^2} - \frac{\partial^2}{\partial y^2} - \frac{\partial^2}{\partial z^2} + m^2 \right) \psi = 0. \tag{6.68}$$

With this equation, Klein and Gordon obtained the wanted symmetry between space and time coordinates and also, since the square root operator was no longer present, the difficulties related to the nonlocality disappeared. Using the d'Alembert operator, we can express the Klein–Gordon equation in its more elegant form:

$$(\Box + m^2)\psi = 0. \tag{6.69}$$

Since $\Box = \partial^\mu \partial_\mu$, the Klein–Gordon equation can also be written as

$$(\partial^\mu \partial_\mu + m^2)\psi = 0. \tag{6.70}$$

So, it is clear that the Klein–Gordon equation takes the same form in any inertial frame of reference, being manifestly invariant.

6.6 Klein–Gordon particle in an electromagnetic field

Let us now consider a Klein–Gordon particle in an electromagnetic field. The canonical momentum $\mathbf{p}$ has to be substituted by the kinetic (or mechanical) momentum $\mathbf{p} - e\mathbf{A}$, where e is the particle charge and $\mathbf{A}$ the vector potential, and E has to be substituted by $E - e\varphi$, where $e\varphi$ is the potential energy (see Eq. (6.60)). Using four-vectors notations, this means

$$\partial_\mu \rightarrow \partial_\mu + ieA_\mu, \tag{6.71}$$

so that

$$[(\partial^\mu + ieA^\mu)(\partial_\mu + ieA_\mu) + m^2]\psi = 0. \tag{6.72}$$

6.7 Nonrelativistic limit of the Klein–Gordon equation

The Schrödinger equation is the nonrelativistic limit of the Klein–Gordon equation. To demonstrate this, let us consider the plane wave:

$$\psi = \exp[i(\mathbf{p} \cdot \mathbf{r} - Et)], \tag{6.73}$$

describing a free particle with energy E and momentum $\mathbf{p}$. Since

$$E^2 - p^2 = m^2, \tag{6.74}$$

using the correspondence rules:

$$E \rightarrow i\frac{\partial}{\partial t}, \tag{6.75}$$

$$\mathbf{p} \rightarrow -i\nabla, \tag{6.76}$$

we immediately obtain the Klein–Gordon Eq. (6.69). It is clear that, applying the d'Alembert operator to a plane-wave packet defined as

$$\Psi = \int A(\mathbf{p}) \exp[i(\mathbf{p} \cdot \mathbf{r} - Et)] d^3p, \tag{6.77}$$

we obtain the same result. As a consequence, it should be now clear that what we will subsequently discuss (which, for the sake of simplicity, we limit to a single plane wave) applies to any plane-wave linear combination or packet.

In the nonrelativistic approximation, the energy T is expressed by

$$T = E - m = \frac{p^2}{2m}. \tag{6.78}$$

If we define the new function:

$$\phi = \exp\left[i\left(\mathbf{p} \cdot \mathbf{r} - \frac{p^2 t}{2m}\right)\right], \tag{6.79}$$

the non-relativistic plane-wave can be expressed by

$$\psi = \exp(-imt)\,\phi\,.$$ (6.80)

Let us apply now the operator $-\Box$ to the nonrelativistic plane wave ψ:

$$-\Box\psi = \left(\nabla^2 - \frac{\partial^2}{\partial t^2}\right)\exp(-imt)\,\phi$$

$$= \exp(-imt)\,\nabla^2\phi - \frac{\partial}{\partial t}\left[-im\,\exp(-imt)\,\phi + \exp(-imt)\frac{\partial\phi}{\partial t}\right]$$

$$= \exp(-imt)\left(\nabla^2\phi + m^2\,\phi + 2im\,\frac{\partial\phi}{\partial t} - \frac{\partial^2\phi}{\partial t^2}\right).$$

Therefore,

$$-\Box\psi = \exp(-imt)\left(\nabla^2 + m^2 + 2im\,\frac{\partial}{\partial t} - \frac{\partial^2}{\partial t^2}\right)\phi\,.$$ (6.81)

Now, let us observe that

$$\frac{\partial^2\phi}{\partial t^2} = \frac{\partial^2}{\partial t^2}\exp\left[i\left(\mathbf{p}\cdot\mathbf{r} - \frac{p^2 t}{2m}\right)\right]$$

$$= \frac{\partial}{\partial t}\left(-i\frac{p^2}{2m}\right)\exp\left[i\left(\mathbf{p}\cdot\mathbf{r} - \frac{p^2 t}{2m}\right)\right]$$

$$= \left(-i\frac{p^2}{2m}\right)\left(-i\frac{p^2}{2m}\right)\phi = -\frac{p^4}{4m^2}\phi\,.$$

Please note that $(p^4/4m^2)\phi = -\partial^2\phi/\partial t^2$ can be neglected because, in the nonrelativistic limit, it is much smaller than $m^2\phi$, so

$$-\Box\psi = \exp(-imt)\left(\nabla^2 + m^2 + 2im\,\frac{\partial}{\partial t}\right)\phi\,.$$ (6.82)

Comparing this equation with the Klein–Gordon equation, we conclude that

$$\left(\nabla^2 + m^2 + 2i\,m\,\frac{\partial}{\partial t}\right)\phi = m^2\phi\,,$$ (6.83)

and, hence,

$$\left(\nabla^2 + 2i\,m\,\frac{\partial}{\partial t}\right)\phi = 0\,.$$ (6.84)

Once extended to the whole packet Ψ, this deduction allows us to conclude that the Schrödinger equation;

$$i\frac{\partial\Psi}{\partial t} = -\frac{1}{2m}\nabla^2\Psi = \frac{p^2}{2m}\Psi,$$ (6.85)

represents the nonrelativistic limit of the Klein–Gordon equation. This result was, of course, expected. The Klein–Gordon equation is the relativistic generalization of the Schrödinger equation, thus the latter has to be the nonrelativistic limit of the Klein–Gordon equation. Wolfgang Pauli, in his course given in Zurich during the academic year 1956–57, used this approach to deduce the Schrödinger equation.

6.8 Difficulties of interpretation

Klein–Gordon equation presents two main difficulties of interpretation. The first one concerns the appearance of negative energy states. In fact, we can recast the Klein–Gordon equation so that

$$\left(H - i\frac{\partial}{\partial t}\right)\left(H + i\frac{\partial}{\partial t}\right)\psi = 0.$$ (6.86)

Indeed, from this equation we obtain

$$
\begin{aligned}
0 &= \left(H^2 - i\frac{\partial}{\partial t}H + iH\frac{\partial}{\partial t} + \frac{\partial^2}{\partial t^2}\right)\psi \\
&= \left(H^2 + \frac{\partial^2}{\partial t^2}\right)\psi = \left(p^2 + m^2 + \frac{\partial^2}{\partial t^2}\right)\psi \\
&= \left(\frac{\partial^2}{\partial t^2} - \nabla^2 + m^2\right)\psi = (\Box + m^2)\psi.
\end{aligned}
$$

It is immediately evident that Eq. (6.86) allows both solutions with positive energy, satisfying the equation:

$$\left(H - i\frac{\partial}{\partial t}\right)\psi = 0$$ (6.87)

and solutions with negative energy, satisfying the equation:

$$\left(H + i\frac{\partial}{\partial t}\right)\psi = 0.$$ (6.88)

This means that, in particular, Eq. (6.86) predicts, for particles at rest, negative masses. In this case, indeed, $\mathbf{p} = 0$, and we have

$$i\frac{\partial\psi}{\partial t} = m\psi,$$ (6.89)

corresponding to the description of particles having positive mass, and

$$i\frac{\partial\psi}{\partial t} = -m\psi, \tag{6.90}$$

corresponding to the description of particles having negative mass. We could think that a possible solution of this difficulty could be to exclude all the solutions with negative energy. Actually, this is not possible when the particles are in the presence of fields. In fact, if there are intense potential energies, then particles can pass from states with positive energy to states with negative energy because of the presence of the fields.

There is another difficulty of interpretation of this first quantum-relativistic equation. It is related to the continuity equation that, with the introduction of relativity, requires that (as we will see shortly) the probability density is not necessarily a positive quantity. It is well known that, using the Schrödinger equation, the probability density is a positive quantity given by

$$\rho = \psi^*\psi.$$

Keep in mind that, with this probability density, the continuity equation:

$$\frac{\partial\rho}{\partial t} + \nabla\cdot\mathbf{j} = 0,$$

is satisfied by the probability current density $\mathbf{j}$ given by

$$\mathbf{j} = \frac{i}{2m}(\psi\nabla\psi^* - \psi^*\nabla\psi).$$

If we consider now the Klein–Gordon equation:

$$\frac{\partial^2\psi}{\partial t^2} = \nabla^2\psi - m^2\psi, \tag{6.91}$$

and its hermitian conjugate:

$$\frac{\partial^2\psi^*}{\partial t^2} = \nabla^2\psi^* - m^2\psi^*, \tag{6.92}$$

we can easily see that continuity equation, Eq. (3.29), is satisfied by the following probability density:

$$\rho = i\left(\psi^*\frac{\partial\psi}{\partial t} - \psi\frac{\partial\psi^*}{\partial t}\right) \tag{6.93}$$

and probability current density:

$$\mathbf{j} = i(\psi\nabla\psi^* - \psi^*\nabla\psi). \tag{6.94}$$

In fact, in this case we have

$$
\begin{aligned}
\nabla\cdot\mathbf{j} &= i\psi\left(\frac{\partial^2\psi^*}{\partial t^2} + m^2\psi^*\right) - i\psi^*\left(\frac{\partial^2\psi}{\partial t^2} + m^2\psi\right) \\
&= i\left[\psi\frac{\partial}{\partial t}\left(\frac{\partial\psi^*}{\partial t}\right) - \psi^*\frac{\partial}{\partial t}\left(\frac{\partial\psi}{\partial t}\right)\right] \\
&= i\left[\frac{\partial}{\partial t}\left(\psi\frac{\partial\psi^*}{\partial t}\right) - \frac{\partial\psi}{\partial t}\frac{\partial\psi^*}{\partial t} - \frac{\partial}{\partial t}\left(\psi^*\frac{\partial\psi}{\partial t}\right) + \frac{\partial\psi^*}{\partial t}\frac{\partial\psi}{\partial t}\right] \\
&= i\frac{\partial}{\partial t}\left(\psi\frac{\partial\psi^*}{\partial t} - \psi^*\frac{\partial\psi}{\partial t}\right) = -\frac{\partial\rho}{\partial t}. \tag{6.95}
\end{aligned}
$$

Note that the factors within the parentheses in Eqs. (6.93) and (6.94) are differences between a complex number and its complex conjugate so, to make the probability density ρ and probability current density $\mathbf{j}$ real, we have introduced the imaginary unit i. Please note that the probability density expressed by Eq. (6.93) is not necessarily positive: It can also be negative or null.

Actually, each relativistic approach to quantum theory must confront similar difficulties, and the reason is that relativity allows the transformation of matter into energy and vice versa. As a consequence, quantum-relativistic theories permit the continuous creation and annihilation of particles.

So, a reinterpretation of the wave function in terms of a field is required. Exactly as the Maxwell equations describe fields, the idea of field is also present in the Klein–Gordon equation.

If we consider particles with null mass, the Klein–Gordon equation becomes

$$\Box\psi = 0. \tag{6.96}$$

This is the d'Alembert equation. As is well known, the d'Alembert equation describes electromagnetic fields in a region of space without charges and currents. In the Lorentz gauge, it also describes the evolution of the electromagnetic potentials. Note that the masses of the particles created and annihilated by the electromagnetic fields, the photons, are null. So, the fact that Eq. (6.69) becomes Eq. (6.96) when the mass is null does not only confirm the correctness of the Klein–Gordon theory but also suggests that the wave function has to be reinterpreted as a field.

Anyway, from the historical point of view, we have to say that the temporary abandonment of the Klein–Gordon equation due to these interpretation difficulties prompted further efforts, allowing the discovery of another very important quantum-relativistic equation: the Dirac equation.

6.9 Spin and gyromagnetic ratio of the electron

Before introducing the Dirac equation, we need to remind our readers that at the dawn of quantum mechanics, although the atomic spectra could be roughly described by Bohr's model, it was clear that experimental data about complex atoms, which were accumulating, required further ideas. There were too many inconsistencies and, in particular, the number of electrons that could occupy a given orbital was not known. Also, the behavior of atoms in the presence of magnetic fields needed to be further explored.

Wolfgang Pauli had demonstrated that Bohr's theory could be used to describe, even if roughly, atoms more complex than hydrogen if the maximum number of electrons that could occupy each individual atomic orbital was equal to two. This is the well known Pauli exclusion principle, whose origin Pauli based on an unspecified "duplicity".

In 1925, two young researchers, Samuel Goudsmit and George Uhlenbeck, had demonstrated that all the problems related to the behavior of atoms in magnetic fields could be solved assuming that all electrons are small magnets, each one with the same magnetic force. According to Goudsmit and Uhlenbeck, the origin of this intrinsic electronic magnetism was the rotational motion of every electron around its axis (the spin). Furthermore, since only two values for the electron spin were permitted by quantum mechanics, the mystery of the Pauli's "duplicity" was then clarified. As intensity of the spin could not be modified and, furthermore, had only two possible orientations, only two states of spin were allowed.

A problem with the Goudsmit and Uhlenbeck hypothesis arose immediately. It concerned the gyromagnetic ratio g of the electron that rules the ratio between the electronic magnetism and its spin. According to the classical methods, g should be equal to 1, while Goudsmit and Uhlenbeck had to postulate that g was equal to 2 to obtain a good agreement with the available experimental data. Even if the situation was not clear, because Goudsmit and Uhlenbeck could not explain the value $g = 2$ of the gyromagnetic factor, their model provided good agreement with the experimental evidence.

In 1928 Paul Dirac proposed using first-order differential equations to reconcile the quantum mechanics of electrons with special relativity. He found out that the minimum number of differential equations needed to obtain this result was four. If two equations can explain the "duplicity" postulated by Pauli, four equations seemed, at a first glance, too many. But, as we will soon see, Dirac's "quadruplicity" explains both the electron spin and the discovery of antimatter.

6.10 Dirac equation

To write his equation, Dirac proposed that the Hamiltonian operator H of a free particle had the following form:

$$H = \sum_{j=1}^{3} \alpha^j p_j + \beta m, \tag{6.97}$$

where the four coefficients α^j ($j = 1, 2, 3$) and β had to be determined assuming that

$$H^2 = p^2 + m^2. \tag{6.98}$$

As a consequence,

$$\left(\alpha^j\right)^2 = 1, \tag{6.99}$$

$$\beta^2 = 1, \tag{6.100}$$

$$\alpha^j \beta + \beta \alpha^j = \{\alpha^j, \beta\} = 0, \tag{6.101}$$

and, for each $j \neq k$,

$$\alpha^j \alpha^k + \alpha^k \alpha^j = \{\alpha^j, \alpha^k\} = 0 \quad (j, k = 1, 2, 3). \tag{6.102}$$

There is only one way to satisfy all these conditions: The quantities α^j and β cannot be numbers. If they were numbers, this system of equations would not have solutions. Actually, α^j and β have to be matrices. Furthermore, since the Hamiltonian operator is hermitian, the matrices α^j and β also have to be hermitian. Note that α^j and β have to be 4×4 matrices (at least). This can be easily demonstrated observing that, since $(\alpha^j)^2 = 1$ and $\beta^2 = 1$, the eigenvalues of α^j and β can only be ± 1. Furthermore, as $\alpha^j = -\beta \alpha^j \beta$ and $\beta = -\alpha^j \beta \alpha^j$, we have

$$\mathrm{Tr}\, \alpha^j = \mathrm{Tr}\, \beta = 0, \tag{6.103}$$

due to the cyclic invariance of the trace. As a consequence, the number of diagonal elements has to be an even number. This even number cannot be 2 because, in the set of hermitian matrices 2×2 represented by the Pauli matrices and the identity, only three of them anticommute. Thus. the smallest matrices satisfying our conditions are 4×4. It is very easy to verify, by direct substitution and taking into account the properties of the Pauli's matrices, that a possible representation of the matrices α^j and β is the following:[3]

$$\alpha^j = \begin{pmatrix} 0 & \sigma^j \\ \sigma^j & 0 \end{pmatrix},$$

(6.104)

$$\beta = \begin{pmatrix} I & 0 \\ 0 & -I \end{pmatrix}.$$

(6.105)

Explicitly expressed, these matrices are:

$$\alpha^1 = \alpha_x = \begin{pmatrix} 0 & 0 & 0 & 1 \\ 0 & 0 & 1 & 0 \\ 0 & 1 & 0 & 0 \\ 1 & 0 & 0 & 0 \end{pmatrix},$$

(6.106)

$$\alpha^2 = \alpha_y = \begin{pmatrix} 0 & 0 & 0 & -i \\ 0 & 0 & i & 0 \\ 0 & -i & 0 & 0 \\ i & 0 & 0 & 0 \end{pmatrix},$$

(6.107)

$$\alpha^3 = \alpha_z = \begin{pmatrix} 0 & 0 & 1 & 0 \\ 0 & 0 & 0 & -1 \\ 1 & 0 & 0 & 0 \\ 0 & -1 & 0 & 0 \end{pmatrix},$$

(6.108)

$$\beta = \begin{pmatrix} 1 & 0 & 0 & 0 \\ 0 & 1 & 0 & 0 \\ 0 & 0 & -1 & 0 \\ 0 & 0 & 0 & -1 \end{pmatrix}.$$

(6.109)

To demonstrate that these matrices satisfy the conditions expressed by Eqs. (6.99), (6.100), (6.101), and (6.102), we can use their representations:

$$(\alpha^j)^2 = \begin{pmatrix} 0 & \sigma^j \\ \sigma^j & 0 \end{pmatrix}\begin{pmatrix} 0 & \sigma^j \\ \sigma^j & 0 \end{pmatrix} = \begin{pmatrix} \sigma^j\sigma^j & 0 \\ 0 & \sigma^j\sigma^j \end{pmatrix}$$

$$= \begin{pmatrix} I & 0 \\ 0 & I \end{pmatrix} = I = 1,$$

$$\beta^2 = \begin{pmatrix} I & 0 \\ 0 & -I \end{pmatrix}\begin{pmatrix} I & 0 \\ 0 & -I \end{pmatrix} = \begin{pmatrix} I & 0 \\ 0 & I \end{pmatrix} = I = 1,$$

---

3 Of course, this is not the only possible representation of the α^j and β matrices.

$$\alpha^j \beta + \beta \alpha^j = \begin{pmatrix} 0 & \sigma^j \\ \sigma^j & 0 \end{pmatrix}\begin{pmatrix} I & 0 \\ 0 & -I \end{pmatrix} + \begin{pmatrix} I & 0 \\ 0 & -I \end{pmatrix}\begin{pmatrix} 0 & \sigma^j \\ \sigma^j & 0 \end{pmatrix}$$

$$= \begin{pmatrix} 0 & -\sigma^j \\ \sigma^j & 0 \end{pmatrix} + \begin{pmatrix} 0 & \sigma^j \\ -\sigma^j & 0 \end{pmatrix} = 0 .$$

As for Eq. (6.102), let us calculate, for example, $\alpha^1 \alpha^2 + \alpha^2 \alpha^1$:

$$\alpha^1 \alpha^2 + \alpha^2 \alpha^1 = \begin{pmatrix} 0 & \sigma^1 \\ \sigma^1 & 0 \end{pmatrix}\begin{pmatrix} 0 & \sigma^2 \\ \sigma^2 & 0 \end{pmatrix} + \begin{pmatrix} 0 & \sigma^2 \\ \sigma^2 & 0 \end{pmatrix}\begin{pmatrix} 0 & \sigma^1 \\ \sigma^1 & 0 \end{pmatrix}$$

$$= \begin{pmatrix} \sigma^1 \sigma^2 & 0 \\ 0 & \sigma^1 \sigma^2 \end{pmatrix} + \begin{pmatrix} \sigma^2 \sigma^1 & 0 \\ 0 & \sigma^2 \sigma^1 \end{pmatrix}$$

$$= \begin{pmatrix} \sigma_x \sigma_y & 0 \\ 0 & \sigma_x \sigma_y \end{pmatrix} + \begin{pmatrix} \sigma_y \sigma_x & 0 \\ 0 & \sigma_y \sigma_x \end{pmatrix}$$

$$= \begin{pmatrix} \sigma_x \sigma_y + \sigma_y \sigma_x & 0 \\ 0 & \sigma_x \sigma_y + \sigma_y \sigma_x \end{pmatrix} = \begin{pmatrix} 0 & 0 \\ 0 & 0 \end{pmatrix} = 0 .$$

It is evident that matrices α^j and β are hermitian because the Pauli matrices are hermitian. The first-order equation we were looking for is then given by:

$$(\boldsymbol{\alpha} \cdot \mathbf{p} + \beta m)\Psi = i\frac{\partial \Psi}{\partial t} , \tag{6.110}$$

where the function Ψ (the so-called "spinor") has four components. This is the Dirac equation describing a free particle.

Note that the Dirac equation does not solve the problem of the solutions with negative energy. For example, if we consider a particle at rest, we see that

$$\beta m \Psi = i\frac{\partial \Psi}{\partial t} , \tag{6.111}$$

or, expressing explicitly the four components of the spinor:

$$\beta m \begin{pmatrix} \Psi_1 \\ \Psi_2 \\ \Psi_3 \\ \Psi_4 \end{pmatrix} = i\frac{\partial}{\partial t}\begin{pmatrix} \Psi_1 \\ \Psi_2 \\ \Psi_3 \\ \Psi_4 \end{pmatrix} , \tag{6.112}$$

i. e.,

$$m \begin{pmatrix} 1 & 0 & 0 & 0 \\ 0 & 1 & 0 & 0 \\ 0 & 0 & -1 & 0 \\ 0 & 0 & 0 & -1 \end{pmatrix}\begin{pmatrix} \Psi_1 \\ \Psi_2 \\ \Psi_3 \\ \Psi_4 \end{pmatrix} = i\frac{\partial}{\partial t}\begin{pmatrix} \Psi_1 \\ \Psi_2 \\ \Psi_3 \\ \Psi_4 \end{pmatrix} . \tag{6.113}$$

Thus,

$$m\Psi_1 = i\frac{\partial \Psi_1}{\partial t}, \tag{6.114}$$

$$m\Psi_2 = i\frac{\partial \Psi_2}{\partial t}, \tag{6.115}$$

$$-m\Psi_3 = i\frac{\partial \Psi_3}{\partial t}, \tag{6.116}$$

$$-m\Psi_4 = i\frac{\partial \Psi_4}{\partial t}. \tag{6.117}$$

While the first two equations describe a particle with positive energy, the last two equations describe a particle with negative energy: If $\Psi_3 \propto \exp(-iEt)$, then

$$- m\exp(-iEt) = i\frac{\partial}{\partial t}\exp(-iEt) = i(-i)E\exp(-iEt) = E\exp(-iEt), \tag{6.118}$$

and, therefore,

$$E = -m.^4 \tag{6.119}$$

We conclude this chapter by mentioning another important set of properties of the α^j matrices, i. e.,

$$\alpha_x\alpha_y = i\sigma_z,$$
$$\alpha_y\alpha_z = i\sigma_x, \tag{6.120}$$
$$\alpha_z\alpha_x = i\sigma_y.$$

The proof is very simple:

$$\alpha_x\alpha_y = \begin{pmatrix} 0 & \sigma_x \\ \sigma_x & 0 \end{pmatrix}\begin{pmatrix} 0 & \sigma_y \\ \sigma_y & 0 \end{pmatrix} = \begin{pmatrix} \sigma_x\sigma_y & 0 \\ 0 & \sigma_x\sigma_y \end{pmatrix}$$
$$= \begin{pmatrix} i\sigma_z & 0 \\ 0 & i\sigma_z \end{pmatrix} = i\sigma_z.$$

---

4 Solutions with negative energy represent positrons, i. e., particles with positive mass m and negative charge $-e$. They are the antiparticles of electrons. Positrons were experimentally discovered in 1932 by C. Anderson. They are today used in many applicative fields, from materials science to medical physics.

Please note that we are using the same symbols σ_x, σ_y, and σ_z to indicate both the 2×2 spin Pauli matrices and the 4 × 4 doubled spin Pauli matrices given by:

$$\sigma_x = \begin{pmatrix} 0 & 1 & 0 & 0 \\ 1 & 0 & 0 & 0 \\ 0 & 0 & 0 & 1 \\ 0 & 0 & 1 & 0 \end{pmatrix}, \tag{6.121}$$

$$\sigma_y = \begin{pmatrix} 0 & -i & 0 & 0 \\ i & 0 & 0 & 0 \\ 0 & 0 & 0 & -i \\ 0 & 0 & i & 0 \end{pmatrix}, \tag{6.122}$$

$$\sigma_z = \begin{pmatrix} 1 & 0 & 0 & 0 \\ 0 & -1 & 0 & 0 \\ 0 & 0 & 1 & 0 \\ 0 & 0 & 0 & -1 \end{pmatrix}. \tag{6.123}$$

The context will always enable the identification of the matrix size.

7 The Dirac equation and electron spin

Electron spin had to be included as an *ad hoc* hypothesis in nonrelativistic quantum mechanics to account for the experimental evidence. The use of the Dirac quantum-relativistic equation, on the other hand, demonstrates that electron spin is a genuinely and authentically relativistic phenomenon. In this chapter, we will show how electron spin naturally appears when we consider the Dirac equation describing an electron in an electromagnetic field [18, 21, 27, 28].

7.1 Manifestly covariant form of the Dirac equation

Let us introduce the Dirac matrices, defined by

$$\gamma^0 \equiv \beta, \tag{7.1}$$

$$\gamma^j \equiv \beta \alpha^j. \tag{7.2}$$

While γ^0 is hermitian (being identical with β), the matrices γ^j are antihermitian. In fact

$$\gamma^{j\dagger} = \left(\beta \alpha^j\right)^\dagger = \alpha^{j\dagger}\beta^\dagger = \alpha^j\beta = -\beta\alpha^j = -\gamma^j. \tag{7.3}$$

The manifestly covariant form of the Dirac equation of the free particle can be easily obtained using the Dirac matrices. Let us introduce the γ^μ four-vector:

$$\gamma^\mu = (\gamma^0, \gamma^1, \gamma^2, \gamma^3) = (\gamma^0, \boldsymbol{\gamma}) \tag{7.4}$$

and remind our readers that the covariant components of the four-vector describing the partial-differentiation operator are

$$\partial_\mu = \left(\frac{\partial}{\partial t}, \frac{\partial}{\partial x}, \frac{\partial}{\partial y}, \frac{\partial}{\partial z}\right) = \left(\frac{\partial}{\partial t}, \boldsymbol{\nabla}\right). \tag{7.5}$$

Let us now introduce the four-vector p^μ defined as

$$p^\mu = (E, \mathbf{p}). \tag{7.6}$$

The correspondence rules:

$$E \rightarrow i\frac{\partial}{\partial t} \tag{7.7}$$

and

$$\mathbf{p} \rightarrow -i\boldsymbol{\nabla}, \tag{7.8}$$

https://doi.org/10.1515/9783110675375-007

can be written as

$$p^\mu \rightarrow i\partial^\mu. \tag{7.9}$$

Let us now write the Dirac equation (6.110) in the form:

$$\left(-i\sum_{j=1}^{3}\alpha^j\partial_j - i\partial_0 + \beta m\right)\Psi = 0, \tag{7.10}$$

and multiply it on the left by β:

$$\left(-i\sum_{j=1}^{3}\beta\alpha^j\partial_j - i\beta\partial_0 + \beta^2 m\right)\Psi = 0. \tag{7.11}$$

So,

$$\left(-i\sum_{j=1}^{3}\gamma^j\partial_j - i\gamma^0\partial_0 + m\right)\Psi = 0. \tag{7.12}$$

This equation can now be rewritten in the manifestly covariant (invariant, in this specific case) form:

$$(i\gamma^\mu\partial_\mu - m)\Psi = 0. \tag{7.13}$$

So, the Dirac equation takes the same form in any inertial frame of reference. Let us now consider a Dirac particle in an electromagnetic field. The canonical momentum $\mathbf{p}$ has to be substituted by the kinetic (or mechanical) momentum $\mathbf{p} - e\mathbf{A}$, where e is the electron charge and $\mathbf{A}$ the vector potential. Furthermore E has to be substituted with $E - e\varphi$, where $e\varphi$ is the potential energy. Then

$$[\boldsymbol{\alpha}\cdot(\mathbf{p} - e\mathbf{A}) + \beta m + e\varphi]\Psi = i\partial_0\Psi. \tag{7.14}$$

Let us now multiply this equation by β on the left and rearrange it to obtain

$$\left[\sum_j\gamma^j(-i\partial_k + eA_k) + \gamma^0(-i\partial_0 + e\varphi) + m\right]\Psi = 0.$$

Recall that $A^\mu \equiv (\varphi, \mathbf{A})$ is the four-potential of the electromagnetic field, so that the Dirac equation becomes

$$[\gamma^\mu(i\partial_\mu - eA_\mu) - m]\Psi = 0. \tag{7.15}$$

7.2 The Dirac equation and spin

7.2.1 Properties of commutators and anti-commutators

In order to discuss how the electron spin appears in the Dirac theory, let us first recall some commutators and anticommutators properties. Let us consider two operators a^μ and b^ν. The commutator $[a^\mu, b^\nu]$ of a^μ e b^ν is given by $a^\mu b^\nu - b^\nu a^\mu$, while the anticommutator $\{a^\mu, b^\nu\}$ is given by $a^\mu b^\nu + b^\nu a^\mu$. From these definitions, these three properties immediately follow:

$$a^\mu b^\nu = \frac{1}{2}[a^\mu, b^\nu] + \frac{1}{2}\{a^\mu, b^\nu\}, \tag{7.16}$$

$$[a^\mu, a^\nu]\{b_\mu, b_\nu\} = 0, \tag{7.17}$$

$$[a^\mu, a^\nu] b_\mu b_\nu = \frac{1}{2}[a^\mu, a^\nu][b_\mu, b_\nu]. \tag{7.18}$$

The demonstration of the first one is quite simple. In fact,

$$\frac{1}{2}[a^\mu, b^\nu] + \frac{1}{2}\{a^\mu, b^\nu\} = \frac{1}{2}(a^\mu b^\nu - b^\nu a^\mu + a^\mu b^\nu + b^\nu a^\mu) = a^\mu b^\nu.$$

As for the second one, let us observe that

$$[a^\mu, a^\nu]\{b_\mu, b_\nu\} = a^\mu a^\nu b_\mu b_\nu + a^\mu a^\nu b_\nu b_\mu - a^\nu a^\mu b_\mu b_\nu - a^\nu a^\mu b_\nu b_\mu.$$

Let us now recall our readers the Einstein convention on repeated indices. According to this convention, we have to sum over the repeated indices, so that

$$a^\nu a^\mu b_\mu b_\nu = \sum_\nu \sum_\mu a^\nu a^\mu b_\mu b_\nu = \sum_\mu \sum_\nu a^\mu a^\nu b_\nu b_\mu = a^\mu a^\nu b_\nu b_\mu,$$

$$a^\nu a^\mu b_\nu b_\mu = \sum_\nu \sum_\mu a^\nu a^\mu b_\nu b_\mu = \sum_\mu \sum_\nu a^\mu a^\nu b_\mu b_\nu = a^\mu a^\nu b_\mu b_\nu$$

and, hence,

$$[a^\mu, a^\nu]\{b_\mu, b_\nu\} = a^\mu a^\nu b_\mu b_\nu + a^\mu a^\nu b_\nu b_\mu - a^\mu a^\nu b_\nu b_\mu - a^\mu a^\nu b_\mu b_\nu = 0.$$

The third property, Eq. (7.18), follows from the first two properties:

$$[a^\mu, a^\nu] b_\mu b_\nu = \frac{1}{2}[a^\mu, a^\nu][b_\mu, b_\nu] + \frac{1}{2}[a^\mu, a^\nu]\{b_\mu, b_\nu\} = \frac{1}{2}[a^\mu, a^\nu][b_\mu, b_\nu].$$

7.2.2 Spin

Let us now multiply Eq. (7.15) on the left by the operator $[-\gamma^\mu(i\partial_\mu - eA_\mu) - m]$ and obtain

$$[\gamma^\mu \gamma^\nu (\partial_\mu + ieA_\mu)(\partial_\nu + ieA_\nu) + m^2]\Psi = 0. \tag{7.19}$$

From the definition of the Dirac matrices, we have

$$\gamma^\mu \gamma^\nu + \gamma^\nu \gamma^\mu \equiv \{\gamma^\mu, \gamma^\nu\} = 2g^{\mu\nu}, \tag{7.20}$$

where $g^{\mu\nu}$ is the metric tensor: $g^{00} = 1$, $g^{\mu\nu} = -1$ for $\mu = \nu = 1, 2, 3$, and $g^{\mu\nu} = 0$ for $\mu \neq \nu$. In fact[1]

$$\gamma^j \gamma^k = \beta\alpha^j\beta\alpha^k = \begin{pmatrix} 0 & \sigma^j \\ -\sigma^j & 0 \end{pmatrix}\begin{pmatrix} 0 & \sigma^k \\ -\sigma^k & 0 \end{pmatrix} = \begin{pmatrix} -\sigma^j\sigma^k & 0 \\ 0 & -\sigma^j\sigma^k \end{pmatrix},$$

$$\gamma^0 \gamma^j = \beta^2\alpha^j = \begin{pmatrix} 0 & \sigma^j \\ \sigma^j & 0 \end{pmatrix},$$

$$\gamma^j \gamma^0 = \beta\alpha^j\beta = -\alpha^j\beta^2 = \begin{pmatrix} 0 & -\sigma^j \\ -\sigma^j & 0 \end{pmatrix},$$

so that

$$\{\gamma^j, \gamma^k\} = -\begin{pmatrix} \sigma^j\sigma^k + \sigma^k\sigma^j & 0 \\ 0 & \sigma^j\sigma^k + \sigma^k\sigma^j \end{pmatrix} = \begin{cases} 0 & \text{if } j \neq k, \\ -2 & \text{if } j = k, \end{cases}$$

$$\{\gamma^0, \gamma^j\} = -\begin{pmatrix} 0 & \sigma^j - \sigma^j \\ \sigma^j - \sigma^j & 0 \end{pmatrix} = 0,$$

$$\{\gamma_0, \gamma_0\} = 2\beta^2 = 2.$$

Using Eqs. (7.16) and (7.20). we obtain

$$\gamma^\mu \gamma^\nu = \frac{1}{2}[\gamma^\mu, \gamma^\nu] + g^{\mu\nu}. \tag{7.21}$$

As a consequence, we can write

$$\gamma^\mu \gamma^\nu (\partial_\mu + ieA_\mu)(\partial_\nu + ieA_\nu)$$
$$= \frac{1}{2}[\gamma^\mu, \gamma^\nu](\partial_\mu + ieA_\mu)(\partial_\nu + ieA_\nu)$$
$$+ (\partial^\nu + ieA^\nu)(\partial_\nu + ieA_\nu),$$

---

1 Please keep in mind that Latin letters $j, k, \ldots$ assume the values 1, 2, and 3.

and, using Eq. (7.18),

$$\gamma^\mu \gamma^\nu (\partial_\mu + ieA_\mu)(\partial_\nu + ieA_\nu)$$

$$= \frac{1}{4}[\gamma^\mu, \gamma^\nu][\partial_\mu + ieA_\mu, \partial_\nu + ieA_\nu]$$

$$+ (\partial^\mu + ieA^\mu)(\partial_\mu + ieA_\mu). \tag{7.22}$$

Let us now introduce the completely contravariant components $S^{\mu\nu}$ of the *spin four-tensor*:

$$S^{\mu\nu} \equiv \frac{i}{4}[\gamma^\mu, \gamma^\nu], \tag{7.23}$$

and observe that

$$[\gamma^\mu \gamma^\nu (\partial_\mu + ieA_\mu)(\partial_\nu + ieA_\nu)] \Psi$$

$$= [(\partial^\mu + ieA^\mu)(\partial_\mu + ieA_\mu) - iS^{\mu\nu} ie(\partial_\mu A_\nu - \partial_\nu A_\mu)] \Psi, \tag{7.24}$$

where we have used the equation:

$$[\partial_\mu + ieA_\mu, \partial_\nu + ieA_\nu] \Psi = ie(\partial_\mu A_\nu - \partial_\nu A_\mu) \Psi. \tag{7.25}$$

Let us now remind our readers that the completely covariant components $F_{\mu\nu}$ of the electromagnetic four-tensor can be expressed, as a function of the components of the four-potential, as

$$F_{\mu\nu} \equiv \partial_\mu A_\nu - \partial_\nu A_\mu, \tag{7.26}$$

and thus

$$[\gamma^\mu \gamma^\nu (\partial_\mu + ieA_\mu)(\partial_\nu + ieA_\nu)] \Psi$$

$$= [(\partial^\mu + ieA^\mu)(\partial_\mu + ieA_\mu) + e S^{\mu\nu} F_{\mu\nu}] \Psi. \tag{7.27}$$

As a consequence, we conclude that the Dirac equation of a charged particle in an electromagnetic field can be expressed in the form:

$$[(\partial^\mu + ieA^\mu)(\partial_\mu + ieA_\mu) + m^2 + e S^{\mu\nu} F_{\mu\nu}] \Psi = 0. \tag{7.28}$$

The physical meaning of this equation can be particularly appreciated by comparing it with the covariant form of the Klein–Gordon equation, describing a charged particle without spin in an electromagnetic field, Eq. (6.72). The Dirac Eq. (7.28) and the Klein–Gordon Eq. (6.72) differ for the term $e S^{\mu\nu} F_{\mu\nu}$. Note that the Klein–Gordon equation describes a particle without spin. The additional term $e S^{\mu\nu} F_{\mu\nu}$ in the Dirac equation represents the coupling of the spin of the electron (or the positron) to the electromagnetic field. So, spin appears in the Dirac theory in a completely natural way. There is no need at all to introduce *ad hoc* hypotheses to include spin in the equations describing electrons and positrons. Actually, spin is a genuinely and authentically relativistic phenomenon.

7.3 The solution of the Dirac equation for the free particle

Let us consider now a free electron (or a free positron). If we factorize the spinor, denoting by $u(\mathbf{p})$ the factor independent of $\mathbf{r}$, the particle is described by the four-component function:

$$\Psi = u(\mathbf{p}) \exp[i(\mathbf{p} \cdot \mathbf{r} - Et)] . \tag{7.29}$$

The Dirac equation describing the functions $u(\mathbf{p})$ takes the form

$$(\boldsymbol{\alpha} \cdot \mathbf{p} + \beta m) u(\mathbf{p}) = E u(\mathbf{p}), \tag{7.30}$$

and this is equivalent to the system:

$$\begin{cases}
(E - m)u_1 - p_z u_3 - (p_x - ip_y)u_4 = 0, \\
(E - m)u_2 - (p_x + ip_y)u_3 + p_z u_4 = 0, \\
-p_z u_1 - (p_x - ip_y)u_2 + (E + m)u_3 = 0, \\
-(p_x + ip_y)u_1 + p_z u_2 + (E + m)u_4 = 0,
\end{cases}$$

where the spinor $u(\mathbf{p})$ has been expressed in term of its four components u_1, u_2, u_3 and u_4:

$$u(\mathbf{p}) = \begin{pmatrix} u_1 \\ u_2 \\ u_3 \\ u_4 \end{pmatrix} . \tag{7.31}$$

To keep the discussion simple we choose, without any loss of generality, the z axis in the direction of $\mathbf{p}$, so as to be able to set $p_x = p_y = 0$ and $p_z = p$. Thus, we have

$$\begin{cases}
(E - m)u_1 - pu_3 = 0, \\
(E - m)u_2 + pu_4 = 0, \\
-pu_1 + (E + m)u_3 = 0, \\
pu_2 + (E + m)u_4 = 0,
\end{cases}$$

or

$$\begin{pmatrix} E - m & 0 & -p & 0 \\ 0 & E - m & 0 & p \\ -p & 0 & E + m & 0 \\ 0 & p & 0 & E + m \end{pmatrix} \begin{pmatrix} u_1 \\ u_2 \\ u_3 \\ u_4 \end{pmatrix} = 0. \tag{7.32}$$

To find nontrivial solutions, we impose that the determinant of the matrix of the coefficients of this system be equal to zero:

$$\begin{vmatrix} E - m & 0 & -p & 0 \\ 0 & E - m & 0 & p \\ -p & 0 & E + m & 0 \\ 0 & p & 0 & E + m \end{vmatrix} = 0. \tag{7.33}$$

In this way, we obtain the eigenvalues of the energy. The eigenvalue equation takes the following simple form:

$$(E^2 - m^2 - p^2)^2 = 0. \tag{7.34}$$

The doubly degenerate eigenvalues of the energy are therefore

$$E_{\pm} = \pm\sqrt{p^2 + m^2}. \tag{7.35}$$

If we denote by E_p the positive eigenvalue,

$$E_p = \sqrt{p^2 + m^2}, \tag{7.36}$$

the two energy eigenvalues can be expressed as

$$E_{\pm} = \pm E_p. \tag{7.37}$$

Let us now consider the spin operator in the direction of $\mathbf{p} = (0, 0, p)$. This is the operator S^{12}. Indeed,

$$\begin{aligned} S^{12} &= \frac{i}{4}[\gamma^1, \gamma^2] = \frac{i}{4}(\beta\alpha^1\beta\alpha^2 - \beta\alpha^2\beta\alpha^1) \\ &= \frac{i}{4}(-\alpha^1\alpha^2 + \alpha^2\alpha^1) = \frac{i}{4}(-\alpha^1\alpha^2 - \alpha^1\alpha^2) \\ &= -i\alpha_x\alpha_y/2 = \sigma_z/2 = S_z. \end{aligned} \tag{7.38}$$

The spin operator in the z direction commutes, of course, with the Hamiltonian operator $H = \alpha_z p + \beta m$. The solutions we are looking for are the eigenvectors common to the Hamiltonian H and the z component of the spin operator $S_z = \sigma_z/2$. Let us first consider the case where $u_1 = 1$ and $u_2 = 0$. After normalizing, using the condition $u^\dagger u = 1$, we find that

$$u_{\uparrow E_+}(p) = \sqrt{\frac{E_p + m}{2E_p}} \begin{pmatrix} 1 \\ 0 \\ \frac{p}{E_p+m} \\ 0 \end{pmatrix}, \qquad (7.39)$$

where $u_{\uparrow E_+}(p)$ is the eigenvector with spin up and positive energy. The other three eigenvectors can be obtained using a similar procedure. The eigenvector with spin down and positive energy is

$$u_{\downarrow E_+}(p) = \sqrt{\frac{E_p + m}{2E_p}} \begin{pmatrix} 0 \\ 1 \\ 0 \\ -\frac{p}{E_p+m} \end{pmatrix}, \qquad (7.40)$$

and the two eigenvectors with negative energy (with spin up and spin down, respectively) are:

$$u_{\uparrow E_-}(p) = \sqrt{\frac{E_p + m}{2E_p}} \begin{pmatrix} -\frac{p}{E_p+m} \\ 0 \\ 1 \\ 0 \end{pmatrix}, \qquad (7.41)$$

$$u_{\downarrow E_-}(p) = \sqrt{\frac{E_p + m}{2E_p}} \begin{pmatrix} 0 \\ \frac{p}{E_p+m} \\ 0 \\ 1 \end{pmatrix}. \qquad (7.42)$$

Let us consider, for example, the positive energy solution corresponding to spin up and calculate the ratio between the u_3 component and the u_1 component:

$$u_3/u_1 = p/(E_p + m). \qquad (7.43)$$

In the nonrelativistic limit, this ratio is of the order of v: Therefore, u_3 is negligible compared to u_1. In other words, in the nonrelativistic limit, the Dirac theory with four-component spinors is reduced to a theory with two-component spinors. We will further discuss, in the next section, this important result, considering the much more interesting case of the nonrelativistic limit for the description of a Dirac particle in an electromagnetic field.

7.4 The Pauli equation

Let us remind our readers that in 1925 the spin model proposed by Goudsmit and Uhlenbeck provided results in agreement with all the experimental evidence about spin, apart from the value of g that had to be equal to 2 instead of 1.

Let us consider an electron in an electromagnetic field. To proceed, let us express the four-component spinor Ψ as

$$\Psi = \begin{pmatrix} \Phi \\ \chi \end{pmatrix}, \tag{7.44}$$

where Φ and χ are two-components spinors. It is easy to see that the Dirac equation can be reformulated into the following pair of equations:

$$\boldsymbol{\sigma} \cdot (\mathbf{p} - e\mathbf{A})\chi = (E - e\varphi - m)\Phi, \tag{7.45}$$

$$\boldsymbol{\sigma} \cdot (\mathbf{p} - e\mathbf{A})\Phi = (E - e\varphi + m)\chi. \tag{7.46}$$

In the last two equations, the components of the three-vector $\boldsymbol{\sigma}$ are the Pauli matrices. From Eq. (7.46) we obtain

$$\chi = \frac{\boldsymbol{\sigma} \cdot (\mathbf{p} - e\mathbf{A})}{(T - e\varphi + 2m)}\Phi, \tag{7.47}$$

where

$$T = E - m. \tag{7.48}$$

Let us now see how Eq. (7.47) changes in the nonrelativistic limit. In this case,

$$e\varphi \ll m \tag{7.49}$$

and

$$T \ll m, \tag{7.50}$$

so that

$$\chi = \frac{\boldsymbol{\sigma} \cdot (\mathbf{p} - e\mathbf{A})}{2m}\Phi. \tag{7.51}$$

Since

$$\frac{\boldsymbol{\sigma} \cdot (\mathbf{p} - e\mathbf{A})}{2m} = \mathcal{O}(v) = \mathcal{O}\left(\frac{v}{c}\right), \tag{7.52}$$

Eq. (7.51) demonstrates that the two "high" components of the four-component spinor are much larger than the two "low" components. In the nonrelativistic limit, χ and Φ are the so-called small and large components, respectively.

From Eqs. (7.45) and (7.51), it follows that

$$\left\{ \frac{1}{2m} [\boldsymbol{\sigma} \cdot (\mathbf{p} - e\mathbf{A})] [\boldsymbol{\sigma} \cdot (\mathbf{p} - e\mathbf{A})] + e\varphi - T \right\} \Phi = 0. \tag{7.53}$$

For any pair of vectors $\mathbf{a}$ and $\mathbf{b}$, with simple algebraic manipulations, it is possible to show that

$$(\boldsymbol{\sigma} \cdot \mathbf{a})(\boldsymbol{\sigma} \cdot \mathbf{b}) = \mathbf{a} \cdot \mathbf{b} + i\boldsymbol{\sigma} \cdot \mathbf{a} \times \mathbf{b}. \tag{7.54}$$

Furthermore, it is also easy to prove that

$$(\mathbf{p} - e\mathbf{A}) \times (\mathbf{p} - e\mathbf{A}) = ie\nabla \times \mathbf{A}. \tag{7.55}$$

Since $\nabla \times \mathbf{A} = \mathcal{H}$, where $\mathcal{H}$ is the magnetic field, we can conclude that

$$(T - e\varphi) \Phi = \left[\frac{(\mathbf{p} - e\mathbf{A})^2}{2m} - \frac{e}{2m} \boldsymbol{\sigma} \cdot \mathcal{H} \right] \Phi. \tag{7.56}$$

The spin operator $\mathbf{S}$ is given by

$$\mathbf{S} = \frac{\boldsymbol{\sigma}}{2}, \tag{7.57}$$

so that Eq. (7.56) can be rewritten as

$$(T - e\varphi) \Phi = \left[\frac{(\mathbf{p} - e\mathbf{A})^2}{2m} - g \frac{e}{2m} \mathbf{S} \cdot \mathcal{H} \right] \Phi, \tag{7.58}$$

where $g = 2$ is the gyromagnetic ratio of the electron. Equation (7.58) is the Pauli equation. g is equal to 2, as correctly predicted by the semiempirical model by Goudsmit and Uhlenbeck which was based on a semiclassical visualization of the electron. This is the most acclaimed success of the Dirac equation. Not only the spin appears as a logical consequence of the simultaneous presence of quantum mechanics and relativity, but the electron gyromagnetic ratio is predicted to be 2—in agreement with the experimental evidence—without introducing any *ad hoc* hypotheses. Indeed, Dirac theory is independent of any visualization of spin, which instead appears as a property of matter that derives directly from the need to merge quantum mechanics and relativity into a single self-consistent theory. The electron, the simplest of particles, must possess this property not by virtue of an intuitive model based on preconceptions borrowed from classical physics (small electrically charged spheres rotating around their axis, as it was in the semiempirical model proposed by Goudsmit and Uhlenbeck) but rather to make its behavior consistent with the fundamental principles of quantum mechanics and relativity. While retaining its most relevant characteristics, like the spin and the correct value of the gyromagnetic ratio, Dirac theory manages, at the same time, to make the intuitive arguments of the model by Goudsmit and Uhlenbeck superfluous.

8 The Dirac theory of atoms

One of the most important applications of the Dirac theory and, in particular, of the Dirac equation in a central field, is the description of the atom. In this chapter, after a discussion about the radial Dirac equation describing an electron in a central field [12], we will use it to investigate the atoms and the energy levels [21, 28]. Additionally, we will show that, in the nonrelativistic limit, the Dirac theory provides exactly the same energy levels predicted by the quantum theory [21, 28].

8.1 Dirac equation in a central field

The Hamiltonian operator of a Dirac particle in a central electromagnetic field is given by

$$H = \boldsymbol{\alpha} \cdot \mathbf{p} + \beta m + V(r), \tag{8.1}$$

where $V(r)$ is the potential energy. Please note that, since

$$\nabla V(r) = \frac{\partial V(r)}{\partial r} \frac{\mathbf{r}}{r}, \tag{8.2}$$

then

$$\mathbf{r} \times \nabla V(r) = 0. \tag{8.3}$$

Let us indicate with $\mathbf{L} = \mathbf{r} \times \mathbf{p}$ the orbital angular momentum of the electron and observe that, due to Eq. (8.3), it commutes with $V(r)$. Indeed, by applying the commutator of $V(r)$ and $\mathbf{L}$ to Ψ, we obtain

$$[V, \mathbf{L}] \Psi = [V, \mathbf{r} \times \mathbf{p}] \Psi = -i[V, \mathbf{r} \times \nabla] \Psi$$
$$= -iV\mathbf{r} \times (\nabla\Psi) + i\mathbf{r} \times \nabla(V\Psi)$$
$$= -iV\mathbf{r} \times (\nabla\Psi) + i\mathbf{r} \times (V\nabla\Psi) + (i\mathbf{r} \times \nabla V)\Psi = 0. \tag{8.4}$$

Let us indicate with $\boldsymbol{\sigma}$ the vector whose components are the Pauli matrices. We wish to demonstrate that the total angular momentum $\mathbf{J}$, defined as

$$\mathbf{J} = \mathbf{L} + \frac{\boldsymbol{\sigma}}{2}, \tag{8.5}$$

is a constant of motion, while neither $\mathbf{L}$ nor $\boldsymbol{\sigma}$ are constants of motion. Let us start calculating the time derivative of the angular momentum $\mathbf{L}$. Since

$$\frac{d\mathbf{L}}{dt} = i[H, \mathbf{L}] = i[\boldsymbol{\alpha} \cdot \mathbf{p} + \beta m + V(r), \mathbf{r} \times \mathbf{p}], \tag{8.6}$$

https://doi.org/10.1515/9783110675375-008

we have, in particular,

$$\frac{dL_z}{dt} = i[\alpha_x p_x + \alpha_y p_y + \alpha_z p_z, x p_y - y p_x] = i\alpha_x [p_x, x] p_y - i\alpha_y [p_y, y] p_x. \qquad (8.7)$$

The commutators $[p_x, x]$, $[p_y, y]$ (and $[p_z, z]$) can be easily calculated keeping in mind that they are operators and applying them to Ψ. For example,

$$[p_x, x]\Psi = [-i\partial_1, x]\Psi = \left[-i\frac{\partial}{\partial_x}, x\right]\Psi$$

$$= -i\frac{\partial}{\partial x}(x\Psi) + ix\frac{\partial\Psi}{\partial x} = -i\Psi - ix\frac{\partial\Psi}{\partial x} + ix\frac{\partial\Psi}{\partial x} = -i\Psi.$$

Thus, we have

$$[p_x, x] = [p_y, y] = [p_z, z] = -i, \qquad (8.8)$$

and, then,

$$\frac{dL_z}{dt} = i\alpha_x(-i)p_y - i\alpha_y(-i)p_x = \alpha_x p_y - \alpha_y p_x = (\boldsymbol{\alpha} \times \mathbf{p})_z. \qquad (8.9)$$

Proceeding in the same way with the components x and y of $\mathbf{L}$, we conclude that

$$\frac{d\mathbf{L}}{dt} = \boldsymbol{\alpha} \times \mathbf{p}. \qquad (8.10)$$

Let us now calculate the time derivative of $\boldsymbol{\sigma}$. Since

$$\frac{d\boldsymbol{\sigma}}{dt} = i[H, \boldsymbol{\sigma}] = i[\boldsymbol{\alpha} \cdot \mathbf{p} + \beta m + V(r), \boldsymbol{\sigma}], \qquad (8.11)$$

we have, in particular,

$$\frac{d\sigma_z}{dt} = i[\boldsymbol{\alpha} \cdot \mathbf{p} + \beta m + V(r), \sigma_z] = i[\boldsymbol{\alpha} \cdot \mathbf{p} + \beta m, -i\alpha_x \alpha_y]. \qquad (8.12)$$

From the properties of the α_x, α_y, α_z, and β matrices, it immediately follows that

$$[\alpha_z, \alpha_x \alpha_y] = 0,$$
$$[\beta, \alpha_x \alpha_y] = 0, \qquad (8.13)$$

so that

$$\frac{d\sigma_z}{dt} = i[\alpha_x p_x + \alpha_y p_y, -i\alpha_x \alpha_y] = [\alpha_x p_x, \alpha_x \alpha_y] + [\alpha_y p_y, \alpha_x \alpha_y]$$

$$= p_x(\alpha_x^2 \alpha_y - \alpha_x \alpha_y \alpha_x) + p_y(\alpha_y \alpha_x \alpha_y - \alpha_x \alpha_y^2)$$

$$= 2(p_x \alpha_y - p_y \alpha_x) = -2(\boldsymbol{\alpha} \times \mathbf{p})_z. \qquad (8.14)$$

As a consequence,

$$\frac{d\boldsymbol{\sigma}}{dt} = -2\,\boldsymbol{\alpha}\times\mathbf{p}. \tag{8.15}$$

Thus,

$$\frac{d\mathbf{L}}{dt} + \frac{1}{2}\frac{d\boldsymbol{\sigma}}{dt} = 0, \tag{8.16}$$

or

$$\frac{d\mathbf{J}}{dt} = 0. \tag{8.17}$$

$\mathbf{J}$ is thus a constant of motion. In order to proceed, we need to express $\boldsymbol{\alpha}\cdot\mathbf{p}$ as a function of the operator $\mathcal{K}$, defined by

$$\mathcal{K} = \beta(1 + \boldsymbol{\sigma}\cdot\mathbf{L}).^{1} \tag{8.18}$$

From the definition of $\mathbf{J}$ we have

$$\mathbf{J}^2 - \mathbf{L}^2 = \frac{3}{4} + \boldsymbol{\sigma}\cdot\mathbf{L}, \tag{8.19}$$

and, as a consequence,

$$\mathcal{K} = \beta\left(\mathbf{J}^2 - \mathbf{L}^2 + \frac{1}{4}\right). \tag{8.20}$$

Since

$$\boldsymbol{\sigma}\cdot\mathbf{L} = \sigma_x L_x + \sigma_y L_y + \sigma_z L_z = \begin{pmatrix} L_z & L_x - iL_y & 0 & 0 \\ L_x + iL_y & -L_z & 0 & 0 \\ 0 & 0 & L_z & L_x - iL_y \\ 0 & 0 & L_x + iL_y & -L_z \end{pmatrix}, \tag{8.21}$$

we have

$$1 + \boldsymbol{\sigma}\cdot\mathbf{L} = \begin{pmatrix} L_z + 1 & L_x - iL_y & 0 & 0 \\ L_x + iL_y & -L_z + 1 & 0 & 0 \\ 0 & 0 & L_z + 1 & L_x - iL_y \\ 0 & 0 & L_x + iL_y & -L_z + 1 \end{pmatrix} = \begin{pmatrix} A & 0 \\ 0 & A \end{pmatrix}, \tag{8.22}$$

---

1 Please note that 1 is the 4×4 identity matrix I, and the three components of $\boldsymbol{\sigma}$ are the doubled Pauli spin matrices. Keep in mind that we use the same symbols σ_x, σ_y, and σ_z to indicate both the 2×2 spin Pauli matrices and the 4×4 doubled spin Pauli matrices. Also, the symbol 1 is used to indicate the number 1, the 2×2 unit matrix, and the 4×4 unit matrix, depending on the context, which always allows identification of the matrix size.

where

$$A = \begin{pmatrix} L_z + 1 & L_x - iL_y \\ L_x + iL_y & -L_z + 1 \end{pmatrix}.$$
(8.23)

Thus,

$$K = \beta(1 + \boldsymbol{\sigma} \cdot \mathbf{L}) = \begin{pmatrix} 1 & 0 \\ 0 & -1 \end{pmatrix} \begin{pmatrix} A & 0 \\ 0 & A \end{pmatrix}$$

$$= \begin{pmatrix} A & 0 \\ 0 & -A \end{pmatrix} = \begin{pmatrix} L_z + 1 & L_x - iL_y & 0 & 0 \\ L_x + iL_y & -L_z + 1 & 0 & 0 \\ 0 & 0 & -L_z - 1 & -L_x + iL_y \\ 0 & 0 & -L_x - iL_y & L_z - 1 \end{pmatrix},$$
(8.24)

$$K\beta = \begin{pmatrix} A & 0 \\ 0 & -A \end{pmatrix} \begin{pmatrix} 1 & 0 \\ 0 & -1 \end{pmatrix} = \begin{pmatrix} A & 0 \\ 0 & A \end{pmatrix} = 1 + \boldsymbol{\sigma} \cdot \mathbf{L},$$
(8.25)

$$\beta K = \begin{pmatrix} 1 & 0 \\ 0 & -1 \end{pmatrix} \begin{pmatrix} A & 0 \\ 0 & -A \end{pmatrix} = \begin{pmatrix} A & 0 \\ 0 & A \end{pmatrix} = 1 + \boldsymbol{\sigma} \cdot \mathbf{L},$$
(8.26)

so that

$$[\beta, K] = 0.$$
(8.27)

It is possible to show that the operator K also commutes with any function of r, with the operator L^2, and with the Hamiltonian H. Let us now introduce the radial-momentum operator p_r, which commutes with K:

$$p_r = -i\frac{1}{r}\frac{\partial}{\partial r} r.$$
(8.28)

Please note that

$$p_r = \frac{\mathbf{r} \cdot \mathbf{p} - i}{r}.$$
(8.29)

This can be easily seen by applying p_r to Ψ:

$$p_r \Psi = \left(-i\frac{1}{r}\frac{\partial}{\partial r} r\right)\Psi = -i\frac{\Psi}{r} - i\frac{1}{r}r\frac{\partial\psi}{\partial r}$$

$$= \frac{1}{r}\left[-i\Psi + r\left(-i\frac{\partial}{\partial r}\right)\Psi\right] = \frac{1}{r}\left[-i\Psi + \mathbf{r}\cdot\left(-i\frac{\partial\Psi}{\partial r}\frac{\mathbf{r}}{r}\right)\right]$$

$$= \frac{1}{r}[-i\Psi + \mathbf{r}\cdot(-i\nabla\Psi)] = \frac{1}{r}(-i + \mathbf{r}\cdot\mathbf{p})\Psi.$$

Let us also introduce the radial component α_r of the operator $\boldsymbol{\alpha}$:

$$\alpha_r = \frac{\boldsymbol{\alpha} \cdot \mathbf{r}}{r}.$$
(8.30)

It anticommutes with β, commutes with $\mathcal{K}$, and has the following property:

$$\alpha_r^2 = 1. \tag{8.31}$$

Indeed,

$$\alpha_r^2 = \frac{(\alpha_x\,x + \alpha_y\,y + \alpha_z\,z)(\alpha_x\,x + \alpha_y\,y + \alpha_z\,z)}{r^2}$$

$$= \frac{\alpha_x^2\,x^2 + \alpha_y^2\,y^2 + \alpha_z^2\,x^2 + \{\alpha_x,\alpha_y\}xy + \{\alpha_y,\alpha_z\}yz + \{\alpha_z,\alpha_x\}zx}{r^2}$$

$$= \frac{x^2 + y^2 + z^2}{r^2} = 1.$$

The same result can be obtained observing that, since for any **a** and **b**,

$$(\boldsymbol{\sigma}\cdot\mathbf{a})(\boldsymbol{\sigma}\cdot\mathbf{b}) = \mathbf{a}\cdot\mathbf{b} + i\boldsymbol{\sigma}\cdot\mathbf{a}\times\mathbf{b},$$

and

$$(\boldsymbol{\alpha}\cdot\mathbf{a})(\boldsymbol{\alpha}\cdot\mathbf{b}) = (\boldsymbol{\sigma}\cdot\mathbf{a})(\boldsymbol{\sigma}\cdot\mathbf{b}),^2$$

we have

$$(\boldsymbol{\alpha}\cdot\mathbf{a})(\boldsymbol{\alpha}\cdot\mathbf{b}) = \mathbf{a}\cdot\mathbf{b} + i\boldsymbol{\sigma}\cdot\mathbf{a}\times\mathbf{b}.$$

Thus,

$$\alpha_r^2 = \frac{(\boldsymbol{\alpha}\cdot\mathbf{r})(\boldsymbol{\alpha}\cdot\mathbf{r})}{r^2} = \frac{\mathbf{r}\cdot\mathbf{r}}{r^2} = 1.$$

Also note that

$$(\boldsymbol{\alpha}\cdot\mathbf{r})(\boldsymbol{\alpha}\cdot\mathbf{p}) = \mathbf{r}\cdot\mathbf{p} + i\boldsymbol{\sigma}\cdot\mathbf{r}\times\mathbf{p} = \mathbf{r}\cdot\mathbf{p} + i\boldsymbol{\sigma}\cdot\mathbf{L}.$$

Since

$$\mathbf{r}\cdot\mathbf{p} = r\,p_r + i$$

and

$$i\beta\mathcal{K} = i\beta^2(1 + \boldsymbol{\sigma}\cdot\mathbf{L}) = i + i\boldsymbol{\sigma}\cdot\mathbf{L},$$

---

2 This identity is left for the reader to be derived as an exercise.

we obtain

$$(\boldsymbol{\alpha} \cdot \mathbf{r})(\boldsymbol{\alpha} \cdot \mathbf{p}) = rp_r + i\beta\mathcal{K}. \tag{8.32}$$

Equation (8.32) is equivalent to

$$\alpha_r(\boldsymbol{\alpha} \cdot \mathbf{p}) = \left(p_r + \frac{i\beta\mathcal{K}}{r} \right), \tag{8.33}$$

or, after multiplying on the left by α_r,

$$(\boldsymbol{\alpha} \cdot \mathbf{p}) = \alpha_r \left(p_r + \frac{i\beta\mathcal{K}}{r} \right). \tag{8.34}$$

As a consequence, the Hamiltonian operator H can be rewritten as

$$H = \alpha_r \left(p_r + \frac{i\beta\mathcal{K}}{r} \right) + \beta m + V(r), \tag{8.35}$$

and the Dirac equation can be expressed by

$$\left[\alpha_r \left(p_r + \frac{i\beta\mathcal{K}}{r} \right) + \beta m + V(r) \right] \Psi = E\Psi. \tag{8.36}$$

8.2 Dirac radial equations

Let us indicate with ζ the eigenvector common to the commuting set of operators β, $\mathcal{K}, L^2$, and J_z, so that

$$\beta\zeta = \zeta, \tag{8.37}$$
$$\mathcal{K}\zeta = -k\zeta, \tag{8.38}$$
$$\mathbf{L}^2\zeta = l(l+1)\zeta, \tag{8.39}$$
$$J_z\zeta = m_j\zeta, \tag{8.40}$$

where $1, -k, l(l+1)$ and m_z are their eigenvalues. Let us now define a new function, η, as

$$\eta = -\alpha_r \zeta. \tag{8.41}$$

It is evident that η has the following properties:

$$\zeta = -\alpha_r \eta, \tag{8.42}$$
$$\beta\eta = -\beta\alpha_r\zeta = \alpha_r\beta\zeta = \alpha_r\zeta = -\eta, \tag{8.43}$$

and

$$\mathcal{K}\eta = -k\eta. \tag{8.44}$$

Let us now linearly combine ζ and η to obtain the spinor we are looking for:

$$\Psi = F(r)\eta + iG(r)\zeta. \tag{8.45}$$

It is easy to see, using this equation and the Dirac equation (8.36), that

$$\alpha_r p_r F(r)\eta = i\left[\frac{dF(r)}{dr} + \frac{F(r)}{r}\right]\zeta, \tag{8.46}$$

$$i\alpha_r p_r G(r)\zeta = -\left[\frac{dG(r)}{dr} + \frac{G(r)}{r}\right]\eta, \tag{8.47}$$

$$\frac{i\alpha_r\beta\mathcal{K}}{r}F(r)\eta = -\frac{i}{r}F(r)k\zeta, \tag{8.48}$$

$$\frac{i\alpha_r\beta\mathcal{K}}{r}iG(r)\zeta = -\frac{1}{r}G(r)k\eta, \tag{8.49}$$

$$\beta m F(r)\eta = -m F(r)\eta, \tag{8.50}$$

$$\beta m iG(r)\zeta = im G(r)\zeta. \tag{8.51}$$

Since ζ and η are linearly independent, we obtain

$$[E + m - V(r)]F(r) + \frac{dG(r)}{dr} + \frac{1+k}{r}G(r) = 0, \tag{8.52}$$

$$-[E - m - V(r)]G(r) + \frac{dF(r)}{dr} + \frac{1-k}{r}F(r) = 0, \tag{8.53}$$

where

$$k = -(l+1), \tag{8.54}$$

when

$$j = l + \frac{1}{2}, \tag{8.55}$$

(spin up) and

$$k = l, \tag{8.56}$$

when

$$j = l - \frac{1}{2} \tag{8.57}$$

(spin down). Indeed, from Eq. (8.20) we obtain

$$-k = j(j+1) - \left(j - \frac{1}{2}\right)\left(j - \frac{1}{2} + 1\right) + \frac{1}{4} = j + \frac{1}{2},$$

for spin up and

$$-k = j(j+1) - \left(j + \frac{1}{2}\right)\left(j + \frac{1}{2} + 1\right) + \frac{1}{4} = -\left(j + \frac{1}{2}\right),$$

for spin down.[3]

Equations (8.52) and (8.53) represent the Dirac radial equations describing an electron in a central field.

8.3 The Dirac theory of one-electron atoms

Let us consider an atom with only one electron and a nucleus with Z protons. If $Z = 1$, then we are dealing with the hydrogen atom. The potential φ has spherical symmetry and is given by the Coulomb law:

$$\varphi(r) = \frac{Ze}{r} . \tag{8.58}$$

so that the potential energy is given by

$$V(r) = -\frac{\kappa}{r} , \tag{8.59}$$

where

$$\kappa = Ze^2 . \tag{8.60}$$

Let us introduce two new variables $u(r)$ and $v(r)$ such that

$$u(r) = -r\, G(r) , \tag{8.61}$$
$$v(r) = r\, F(r) , \tag{8.62}$$

so that the Dirac radial equations become

$$\frac{du}{dr} + \frac{k}{r} u(r) - \left[E + m + \frac{\kappa}{r}\right] v(r) = 0 , \tag{8.63}$$

$$\frac{dv}{dr} - \frac{k}{r} v(r) + \left[E - m + \frac{\kappa}{r}\right] u(r) = 0 . \tag{8.64}$$

---

3 Please note that, since $j = 1/2, 3/2, 5/2, \ldots$, it follows that $k = \pm 1, \pm 2, \pm 3, \ldots$.

Let us define two new variables ϕ_1 and ϕ_2 such that

$$\phi_1 + \phi_2 = \frac{\exp(\lambda r)}{\sqrt{m + E}} u(r), \tag{8.65}$$

$$\phi_1 - \phi_2 = \frac{\exp(\lambda r)}{\sqrt{m - E}} v(r), \tag{8.66}$$

where

$$\lambda = \sqrt{m^2 - E^2}. \tag{8.67}$$

With these definitions, we have

$$\frac{d\phi_1}{dr} + \frac{d\phi_2}{dr} = \left(\lambda - \frac{k}{r}\right)(\phi_1 + \phi_2)$$

$$+ \left(E + m + \frac{\kappa}{r}\right)\sqrt{\frac{m - E}{m + E}}(\phi_1 - \phi_2), \tag{8.68}$$

$$\frac{d\phi_1}{dr} - \frac{d\phi_2}{dr} = \left(\lambda + \frac{k}{r}\right)(\phi_1 - \phi_2)$$

$$- \left(E - m + \frac{\kappa}{r}\right)\sqrt{\frac{m + E}{m - E}}(\phi_1 + \phi_2). \tag{8.69}$$

Adding these two equations, we obtain

$$\frac{d\phi_1}{dr} = 2\lambda\phi_1 - \frac{k}{r}\phi_2 - \frac{E\kappa}{\lambda r}\phi_1 - \frac{m\kappa}{\lambda r}\phi_2, \tag{8.70}$$

while, subtracting Eqs. (8.69) from (8.68), we have

$$\frac{d\phi_2}{dr} = -\frac{k}{r}\phi_1 + \frac{m\kappa}{\lambda r}\phi_1 + \frac{E\kappa}{\lambda r}\phi_2. \tag{8.71}$$

These equations become more manageable by introducing the new variable

$$\rho = 2\lambda r. \tag{8.72}$$

Expressed as a function of ρ, they become

$$\frac{d\phi_1}{d\rho} = \left(1 - \frac{E\kappa}{\rho\lambda}\right)\phi_1 - \left(\frac{k}{\rho} + \frac{m\kappa}{\rho\lambda}\right)\phi_2, \tag{8.73}$$

$$\frac{d\phi_2}{d\rho} = \frac{E\kappa}{\rho\lambda}\phi_2 - \left(\frac{k}{\rho} - \frac{m\kappa}{\rho\lambda}\right)\phi_1. \tag{8.74}$$

To proceed and solve this system of differential equations, let us expand ϕ_1 and ϕ_2 in a power series:

$$\phi_1(\rho) = \sum_{j=0}^{\infty} A_j \rho^{j+s}, \tag{8.75}$$

$$\phi_2(\rho) = \sum_{j=0}^{\infty} B_j \rho^{j+s}. \tag{8.76}$$

We will determine the value of s by requiring that these two power series satisfy the differential equations (8.73) and (8.74). Substituting the power series (8.75) and (8.76) in Eqs. (8.73) and (8.74), we obtain the following recursive relationships:

$$(s+j) A_j - A_{j-1} + \frac{E\kappa}{\lambda} A_j + \left(k + \frac{m\kappa}{\lambda} \right) B_j = 0, \tag{8.77}$$

$$(s+j) B_j + \left(k - \frac{m\kappa}{\lambda} \right) A_j - \frac{E\kappa}{\lambda} B_j = 0. \tag{8.78}$$

Let us now consider the case $j = 0$. Since $A_{-1} = 0$, we have

$$sA_0 + \frac{E\kappa}{\lambda} A_0 + \left(k + \frac{m\kappa}{\lambda} \right) B_0 = 0, \tag{8.79}$$

$$sB_0 + \left(k - \frac{m\kappa}{\lambda} \right) A_0 - \frac{E\kappa}{\lambda} B_0 = 0. \tag{8.80}$$

Let us now rearrange these equations to obtain the following homogeneous system of linear equations:

$$\begin{cases} (s + \frac{E\kappa}{\lambda}) A_0 + (k + \frac{m\kappa}{\lambda}) B_0 = 0, \\ (k - \frac{m\kappa}{\lambda}) A_0 + (s - \frac{E\kappa}{\lambda}) B_0 = 0. \end{cases} \tag{8.81}$$

Such a homogeneous system admits nontrivial solutions only if the determinant of the matrix of coefficients is null:

$$\begin{vmatrix} s + E\kappa/\lambda & k + m\kappa/\lambda \\ k - m\kappa/\lambda & s - E\kappa/\lambda \end{vmatrix} = 0. \tag{8.82}$$

This means that

$$s^2 - \frac{E^2\kappa^2}{\lambda^2} - \left(k^2 - \frac{m^2\kappa^2}{\lambda^2} \right) = 0,$$

and therefore, since

$$\lambda^2 = m^2 - E^2,$$

we conclude that

$$s = \pm \sqrt{k^2 - \kappa^2}. \tag{8.83}$$

To obtain square-integrable wave functions,

$$s = \sqrt{k^2 - \kappa^2}. \tag{8.84}$$

Let us consider now Eqs. (8.77) and (8.78). From Eq. (8.78), in particular, we obtain

$$\frac{B_j}{A_j} = \frac{k - m\kappa/\lambda}{\delta - j}, \tag{8.85}$$

where

$$\delta = E\kappa/\lambda - s. \tag{8.86}$$

Let us insert Eq. (8.85) into Eq. (8.77). By simple algebraic manipulations, we obtain

$$A_j = \frac{j - \delta}{(2s + j)j} A_{j-1}. \tag{8.87}$$

Similarly, we have

$$B_j = \frac{j - 1 - \delta}{(2s + j)j} B_{j-1}. \tag{8.88}$$

Recursively applying these two relationships, we can express A_j as a function of A_0 and B_j as a function of B_0. Taking into account the equation:

$$B_0 = \frac{k - m\kappa/\lambda}{\delta} A_0, \tag{8.89}$$

B_j can also be expressed as a function of A_0. Thus, as can be easily verified using the recursive relationships, the functions $\phi_1(\rho)$ and $\phi_2(\rho)$ of the atoms with one electrons are

$$\phi_1(\rho) = A_0 \rho^s M(1 - \delta, 2s + 1, \rho), \tag{8.90}$$

$$\phi_2(\rho) = A_0 \frac{k - m\kappa/\lambda}{\delta} \rho^s M(-\delta, 2s + 1, \rho), \tag{8.91}$$

where $M = M(\alpha, \beta, \rho)$ is the confluent hypergeometric function:

$$M(\alpha, \beta, \rho) = 1 + \frac{\alpha}{\beta}\rho + \frac{\alpha(\alpha + 1)}{2!\,\beta(\beta + 1)}\rho^2 + \cdots$$
$$+ \frac{\alpha(\alpha + 1)(\alpha + 2) \cdots (\alpha + j - 1)}{j!\,\beta(\beta + 1)(\beta + 2) \cdots (\beta + j - 1)}\rho^j + \cdots. \tag{8.92}$$

Note that, in particular, for $\alpha = 1 - \delta$ and $\beta = 2s + 1$,

$$M(1 - \delta, 2s + 1, \rho) = 1 + \frac{1 - \delta}{2s + 1}\rho + \cdots + \frac{(1 - \delta)(2 - \delta) \cdots (j - \delta)}{j!(2s + 1)(2s + 2) \cdots (2s + j)}\rho^j + \cdots, \quad (8.93)$$

while, for $\alpha = -\delta$ and $\beta = 2s + 1$,

$$M(-\delta, 2s+1, \rho) = 1 - \frac{\delta}{2s + 1}\rho + \cdots + (-1)^j \frac{\delta(\delta - 1) \cdots (\delta - j + 1)}{j!(2s + 1)(2s + 2) \cdots (2s + j)}\rho^j + \cdots. \quad (8.94)$$

Once the functions $\phi_1(\rho)$ and $\phi_2(\rho)$ are known, simple manipulations make it possible to obtain the form of the wave functions $F(r)$ and $G(r)$.

8.4 The Dirac theory of one-electron atom energy levels

Let us now calculate the energy levels. Since

$$\delta = \frac{E\kappa}{\lambda} - s = \frac{E\kappa}{\sqrt{m^2 - E^2}} - s, \quad (8.95)$$

we have

$$\sqrt{m^2 - E^2}\,\delta = E\kappa - s\sqrt{m^2 - E^2}, \quad (8.96)$$

and, as a consequence,

$$\begin{aligned}
E^2 &= m^2\,\frac{(\delta + s)^2}{(\delta + s)^2 + \kappa^2} \\
&= m^2\left[\frac{(\delta + s)^2 + \kappa^2}{(\delta + s)^2}\right]^{-1} \\
&= m^2\left[1 + \frac{\kappa^2}{(\delta + s)^2}\right]^{-1}.
\end{aligned}$$

Let us remind our readers that $|k| = l+1$ in the case of spin up, i. e. $j = l+1/2$, and $|k| = l$ in case of spin down, i. e. $j = l - 1/2$, and introduce the principal quantum number n

$$n = \delta + |k|. \quad (8.97)$$

The energy levels of the atom with only one electron are thus given by

$$E = m\left[1 + \frac{\kappa^2}{(n - |k| + s)^2}\right]^{-1/2}. \quad (8.98)$$

Let us consider the nonrelativistic limit. Since

$$s = \sqrt{k^2 - \kappa^2} = |k|\sqrt{1 - \frac{\kappa^2}{k^2}}, \tag{8.99}$$

s, in this limit, can be approximated as

$$s \approx |k|\left(1 - \frac{\kappa^2}{2k^2}\right). \tag{8.100}$$

As a consequence,

$$\begin{aligned} E &\approx m\left[1 + \frac{\kappa^2}{(n - |k| + |k| - \frac{\kappa^2}{2|k|})^2}\right]^{-1/2} \\ &\approx m\left(1 + \frac{\kappa^2}{n^2}\right)^{-1/2} \\ &\approx m\left(1 - \frac{\kappa^2}{2n^2}\right). \end{aligned}$$

As $\kappa = Z e^2$, once the rest energy m is substracted, we find the equation describing the nonrelativistic one-electron atom energy levels:

$$E - m \approx -\frac{Z^2 e^4 m}{2 n^2}, \tag{8.101}$$

with $n = 1, 2, 3, 4, \ldots$.

8.5 Many-electron atoms

8.5.1 Screening function

As we know, assuming spherical symmetry, the screening function is the ratio between the electrostatic potential experienced by an electron at a distance r from the nucleus (which depends on the screening effect due to the presence of the cloud of atomic electrons) and the electrostatic potential of the bare nucleus. In the literature many approximate analytical screening functions have been suggested, based on the solution of the Schrödinger equation.

In Dirac–Hartree–Fock theory, instead of the Schrödinger equation the Dirac equation is used to calculate one-electron orbitals. In this way all the relativistic effects are included on the one-electron orbitals and binding energies. Following the Slater approximation, to deal with the exchange effects, Dirac–Hartree–Fock–Slater self-consistent fields can be calculated to obtain the atomic screening functions and thus provide the accurate analytical potentials.

Let us consider an electron in a central potential, and let us indicate with $V(r)$ the potential energy:

$$V(r) = -e\varphi(r), \tag{8.102}$$

where $\varphi(r)$ is the central potential. For atoms with only one electron, the potential is given by the Coulomb law [see Eq. (8.58)]:

$$\varphi(r) = \frac{Ze}{r}.$$

We already know that in the case of atoms with many electrons we need to take into account the effect of the screening of the atomic electrons on the Coulomb potential. A possible solution is the introduction of a screening function $\xi(r)$ so that

$$\varphi(r) = \frac{Ze}{r} \xi(r). \tag{8.103}$$

A superimposition of Yukawa potentials represents a very good approximation of the screening function. The screening function is, in this case, described by the following equation:

$$\xi(r) = \sum_i \gamma_i \exp[-\lambda_i r]. \tag{8.104}$$

The values of the parameters γ_i ($\sum_i \gamma_i = 1$) and λ_i are determined to obtain the best fit with Hartree–Fock or, even better, with Dirac–Hartree–Fock self-contained calculations, so that the potential energy of an electron in the central potential due to a many-electron atom can be expressed as (see Cox and Bonham [7] and Salvat et al. [24])

$$V(r) = -\frac{Ze^2}{r} \sum_i \gamma_i \exp[-\lambda_i r]. \tag{8.105}$$

8.5.2 Corrections to the electrostatic potential

Exchange effects
Furthermore, we also need to take into account the exchange of the incident electron with an atomic electron. This phenomenon, known as the exchange effect, can be conveniently taken into consideration by adding an exchange potential to the potential calculated by using the Dirac–Hartree–Fock theory.

The Furness and McCarthy potential V_{ex} (see Furness and McCarthy [13]), even if nonrelativistic, can be used with the Dirac theory because exchange is only a small correction to the electrostatic potential (see Salvat and Mayol [25]). It is given by:

$$V_{ex} = \frac{1}{2}(E - V) - \frac{1}{2}\sqrt{(E - V)^2 + 4\pi a_0 e^4 \rho},$$ (8.106)

where E is the electron kinetic energy, e the electron charge, a_0 the Bohr radius, and ρ the atomic electron density (obtained from the Poisson equation).

Polarization of the electron cloud of the target atom

Another phenomenon that should be taken into account for the description of the atomic potential is the polarization of the electron cloud of the target atom caused by the passage of relatively slow electrons. If the electron is far from the atom, the Buckingham potential V_p represents a good approximation of the effect of polarization of the electron cloud. It is given by

$$V_p = -\frac{\alpha_d e^2}{2(r^2 + d^2)^2},$$ (8.107)

where α_d is the atomic dipole polarizability. Please note that the parameter d avoids the divergence of the Buckingham potential at $r = 0$. According to Salvat et al. [23], it can be calculated by $d^4 = 0.5\,\alpha_d\,a_0\,Z^{-1/3}\,b^2$, where

$$b^2 = \max[(E-50\,\text{eV})/(16\,\text{eV}), 1].$$ (8.108)

Other phenomena

For a more complete description of the aforementioned and other phenomena (such as correlation potential, solid-state effects in the case atoms are bound in a solid, and loss of particles from the elastic channel to the inelastic channels), please see Salvat et al. [23].

9 Relativistic partial wave expansion method

In this chapter, the quantum relativistic theory of the elastic scattering of electrons by atoms will be described [2, 3, 5, 8–10, 12, 14, 17–19, 23, 25]. After a discussion about the scattering amplitudes and their relationship with the differential elastic scattering cross-section, a procedure will be provided to numerically calculate the phase shifts, i. e., the fundamental ingredients to compute scattering amplitudes and elastic scattering cross-sections.

9.1 Scattering amplitudes

9.1.1 The fundamental equation

The fundamental equation of relativistic quantum mechanics is the Dirac equation. The wave function is a four-component spinor. The asymptotic forms of the four components of the scattered wave are

$$\Psi_j \underset{r\to\infty}{\sim} a_j \exp(iKz) + b_j(\vartheta, \varphi)\frac{\exp(iKr)}{r} . \tag{9.1}$$

In this chapter, we will indicate with K the relativistic wave number of the projectile. If the spin is parallel to the direction of incidence (spin up), $a_1 = 1$, $a_2 = 0$, $b_1 = f^+(\vartheta, \varphi)$, $b_2 = g^+(\vartheta, \varphi)$, where f^+ and g^+ are two scattering amplitudes. In this case, the asymptotic behavior of the two "high" components of the four-component spinor, Ψ_1 and Ψ_2, is described by the following equations:

$$\Psi_1 \underset{r\to\infty}{\sim} \exp(iKz) + f^+(\vartheta, \varphi)\frac{\exp(iKr)}{r} , \tag{9.2}$$

$$\Psi_2 \underset{r\to\infty}{\sim} g^+(\vartheta, \varphi)\frac{\exp(iKr)}{r} . \tag{9.3}$$

The case of spin antiparallel to the direction of incidence (spin down) corresponds to $a_1 = 0$, $a_2 = 1$, $b_1 = g^-(\vartheta, \varphi)$, $b_2 = f^-(\vartheta, \varphi)$, and the asymptotic behaviour is, in this second case, described by the following equations:

$$\Psi_1 \underset{r\to\infty}{\sim} g^-(\vartheta, \varphi)\frac{\exp(iKr)}{r} , \tag{9.4}$$

$$\Psi_2 \underset{r\to\infty}{\sim} \exp(iKz) + f^-(\vartheta, \varphi)\frac{\exp(iKr)}{r} . \tag{9.5}$$

The differential elastic scattering cross-section can be calculated as

$$\frac{d\sigma}{d\Omega} = \frac{\sum_{j=1}^{4} |b_j|^2}{\sum_{j=1}^{4} |a_j|^2} . \tag{9.6}$$

https://doi.org/10.1515/9783110675375-009

Asymptotically, the scattered wave is made up of plane waves proceeding from the center in various directions, and the coefficients of the solutions for a plane wave are not all independent. From Eqs. (7.39) and (7.40), we see that

$$\frac{|b_3|}{|b_1|} = \frac{|b_4|}{|b_2|} = \frac{|a_3|}{|a_1|} = \frac{|a_4|}{|a_2|} = \frac{p}{E_p + m}, \tag{9.7}$$

so that

$$\frac{d\sigma}{d\Omega} = \frac{|b_1|^2 + |b_2|^2}{|a_1|^2 + |a_2|^2}. \tag{9.8}$$

We know, from the previous chapter, that the Dirac equations for an electron in a central field are given by

$$[E + m - V(r)]F_l^\pm(r) + \frac{dG_l^\pm(r)}{dr} + \frac{1 + k}{r}G_l^\pm(r) = 0, \tag{9.9}$$

$$-[E - m - V(r)]G_l^\pm(r) + \frac{dF_l^\pm(r)}{dr} + \frac{1 - k}{r}F_l^\pm(r) = 0. \tag{9.10}$$

The superscript "+" refers to the electrons with spin up ($k = -l - 1$) while "−" refers to electrons with spin down ($k = l$). Let us now introduce the following new variables:

$$\mu(r) \equiv E + m - V(r), \tag{9.11}$$

$$v(r) \equiv E - m - V(r), \tag{9.12}$$

and let us indicate with μ' the derivative of μ with respect to r:

$$\mu' = \frac{d\mu}{dr}. \tag{9.13}$$

From Eq. (9.9) we obtain

$$F_l^\pm(r) = -\frac{1}{\mu}\left(\frac{dG_l^\pm}{dr} + \frac{1 + k}{r}G_l^\pm\right) \tag{9.14}$$

and

$$\begin{aligned}\frac{dF_l^\pm}{dr} &= \frac{\mu'}{\mu^2}\left(\frac{dG_l^\pm}{dr} + \frac{1 + k}{r}G_l^\pm\right) \\ &\quad - \frac{1}{\mu}\left(\frac{d^2G_l^\pm}{dr^2} + \frac{1 + k}{r}\frac{dG_l^\pm}{dr} - \frac{1 + k}{r^2}G_l^\pm\right).\end{aligned} \tag{9.15}$$

Using now Eq. (9.10), after a few simple algebraic manipulations, we obtain:

$$\frac{d^2G_l^\pm}{dr^2} + \left(\frac{2}{r} - \frac{\mu'}{\mu}\right)\frac{dG_l^\pm}{dr} + \left(\mu v - \frac{k(k+1)}{r^2} - \frac{1 + k}{r}\frac{\mu'}{\mu}\right)G_l^\pm = 0. \tag{9.16}$$

9.1.2 Effective Dirac potential

Please note that, from

$$K^2 = E^2 - m^2,$$ (9.17)

we see that

$$\mu v = K^2 - 2EV + V^2.$$ (9.18)

Let us now introduce the function $\mathcal{G}_l^{\pm}$:

$$\mathcal{G}_l^{\pm} \equiv \frac{r}{\mu^{1/2}} G_l^{\pm},$$ (9.19)

and define the *effective Dirac potential* $U_l^{\pm}(r)$:

$$-U_l^{\pm}(r) = -2EV + V^2 - \frac{k}{r}\frac{\mu'}{\mu} + \frac{1}{2}\frac{\mu''}{\mu} - \frac{3}{4}\frac{\mu'^2}{\mu^2}.$$ (9.20)

From Eq. (9.16) it follows that:

$$\left[\frac{d^2}{dr^2} - \frac{k(k+1)}{r^2} + K^2 - U_l^{\pm}(r)\right]\mathcal{G}_l^{\pm} = 0.$$ (9.21)

For large values of r, $\mathcal{G}_l^{\pm}$ is essentially sinusoidal. In fact, in this case, $V(r) \approx 0$ and since, in the same limit, $\mu \approx E + m$, it does not depend on r. Then, $U_l^{\pm} \approx 0$. As a consequence, when $r \to \infty$, we have

$$\left[\frac{d^2}{dr^2} - \frac{k(k+1)}{r^2} + K^2\right]\mathcal{G}_l^{\pm} = 0.$$ (9.22)

Please note that $k(k+1) = l(l+1)$. In fact, when spin is down $k = l$, while when spin is up

$$k(k+1) = (-l-1)(-l-1+1) = -(l+1)(-l) = l(l+1).$$

So, when $r \to \infty$, we have

$$\left[\frac{d^2}{dr^2} - \frac{l(l+1)}{r^2} + K^2\right]\mathcal{G}_l^{\pm} = 0.$$ (9.23)

Since $Krj_l(Kr)$ and $Krn_l(Kr)$ are solutions to this equation, we conclude that, for $r \to \infty$, $G_l^{\pm}$ is a linear combination of $j_l(Kr)$ and $n_l(Kr)$.

9.1.3 Phase shifts

We have learned that, when r is large enough, $V(r)$ is negligible, $U_l^\pm$ is negligible as well, and the solution to Eq. (9.22) is therefore a linear combination of the regular and irregular spherical Bessel functions multiplied by Kr. From

$$\mathcal{G}_l^\pm = (r/\mu^{1/2}) G_l^\pm$$

it follows that we can write

$$G_l^\pm \underset{r\to\infty}{\sim} j_l(Kr)\cos\eta_l^\pm - n_l(Kr)\sin\eta_l^\pm, \tag{9.24}$$

where $\eta_l^\pm$ are constants to be determined (the phase shifts). Taking into account the asymptotic behaviour of the Bessel functions,

$$j_l(Kr) \underset{r\to\infty}{\sim} \frac{1}{Kr}\sin\left(Kr - \frac{l\pi}{2}\right), \tag{9.25}$$

$$n_l(Kr) \underset{r\to\infty}{\sim} -\frac{1}{Kr}\cos\left(Kr - \frac{l\pi}{2}\right), \tag{9.26}$$

we can therefore write

$$G_l^\pm \underset{r\to\infty}{\sim} \frac{1}{Kr}\sin\left(Kr - \frac{l\pi}{2}\right)\cos\eta_l^\pm + \frac{1}{Kr}\cos\left(Kr - \frac{l\pi}{2}\right)\sin\eta_l^\pm. \tag{9.27}$$

As a consequence,

$$G_l^+ \underset{r\to\infty}{\sim} \frac{1}{Kr}\sin\left(Kr - \frac{l\pi}{2} + \eta_l^+\right), \tag{9.28}$$

and

$$G_l^- \underset{r\to\infty}{\sim} \frac{1}{Kr}\sin\left(Kr - \frac{l\pi}{2} + \eta_l^-\right). \tag{9.29}$$

The phase shifts $\eta_l^\pm$ represent the effect of the potential $V(r)$ on the phases of the scattered waves.

9.1.4 Scattering amplitudes

Let us expand now Ψ_1 and Ψ_2 in spherical harmonics:

$$\Psi_1 = \sum_{l=0}^{\infty}[A_l G_l^+ + B_l G_l^-]P_l(\cos\vartheta), \tag{9.30}$$

$$\Psi_2 = \sum_{l=1}^{\infty} [C_l G_l^+ + D_l G_l^-] P_l^1(\cos\vartheta) \exp(i\varphi). \tag{9.31}$$

The coefficients A_l, B_l, C_l, and D_l can be determined by considering the asymptotic behaviors of the functions. Let us begin with the function Ψ_1:

$$\Psi_1 - \exp(iKz) = \sum_{l=0}^{\infty} [A_l G_l^+ + B_l G_l^- - (2l+1)i^l j_l(Kr)] P_l(\cos\vartheta). \tag{9.32}$$

Since

$$\Psi_1 - \exp(iKz) \underset{r\to\infty}{\sim} \frac{\exp(iKr)}{r} f^+(\vartheta,\varphi), \tag{9.33}$$

we have:

$$\frac{1}{Kr} \sum_{l=0}^{\infty} \left[A_l \sin\left(Kr - \frac{l\pi}{2} + \eta_l^+ \right) + B_l \sin\left(Kr - \frac{l\pi}{2} + \eta_l^- \right) \right.$$
$$\left. - (2l+1)i^l \sin\left(Kr - \frac{l\pi}{2} \right) \right] P_l(\cos\vartheta)$$
$$= \frac{\exp(iKr)}{r} f^+(\vartheta,\varphi). \tag{9.34}$$

Therefore,

$$\frac{\exp(iKr)}{2iKr} \sum_{l=0}^{\infty} \exp\left(-i\frac{l\pi}{2} \right)$$
$$\times [A_l \exp(i\eta_l^+) + B_l \exp(i\eta_l^-) - (2l+1)i^l] P_l(\cos\vartheta)$$
$$- \frac{\exp(-iKr)}{2iKr} \sum_{l=0}^{\infty} \exp\left(i\frac{l\pi}{2} \right)$$
$$\times [A_l \exp(-i\eta_l^+) + B_l \exp(-i\eta_l^-) - (2l+1)i^l] P_l(\cos\vartheta)$$
$$= \frac{\exp(iKr)}{r} f^+(\vartheta,\varphi). \tag{9.35}$$

To satisfy the asymptotic conditions, the coefficient of

$$-\frac{\exp(-iKr)}{2iKr}$$

must be null, so that we have

$$A_l \exp(-i\eta_l^+) + B_l \exp(-i\eta_l^-) = (2l+1)i^l. \tag{9.36}$$

With the choices

$$A_l = (l + 1)i^l \exp(i\eta_l^+),$$ (9.37)

$$B_l = li^l \exp(i\eta_l^-),$$ (9.38)

Eq. (9.36) is satisfied. Proceeding in a similar way for the Ψ_2 function, since

$$\Psi_2 \underset{r\to\infty}{\sim} \frac{\exp(iKr)}{r} g^+(\vartheta, \varphi),$$ (9.39)

we can write

$$\sum_{l=1}^{\infty} [C_l G_l^+ + D_l G_l^-] P_l^1(\cos\vartheta) \exp(i\varphi) = \frac{\exp(iKr)}{r} g^+(\vartheta, \varphi),$$ (9.40)

obtaining

$$C_l \exp(-i\eta_l^+) + D_l \exp(-i\eta_l^-) = 0.$$ (9.41)

This equation is satisfied if

$$C_l = -i^l \exp(i\eta_l^+),$$ (9.42)

$$D_l = i^l \exp(i\eta_l^-).$$ (9.43)

In conclusion, for electrons with spins parallel to the direction of incidence (spin up),

$$\Psi_1 = \sum_{l=0}^{\infty} [(l + 1) \exp(i\eta_l^+)G_l^+ + l\exp(i\eta_l^-)G_l^-]i^l P_l(\cos\vartheta),$$ (9.44)

$$\Psi_2 = \sum_{l=1}^{\infty} [\exp(i\eta_l^-)G_l^- - \exp(i\eta_l^+)G_l^+]i^l P_l^1(\cos\vartheta) \exp(i\varphi),$$ (9.45)

so that, using Eq. (9.35),

$$f^+(\vartheta, \varphi) = f^+(\vartheta)$$

$$= \frac{1}{2iK} \sum_{l=0}^{\infty} \{(l + 1) [\exp(2i\eta_l^+) - 1]$$

$$+ l [\exp(2i\eta_l^-) - 1]\} P_l(\cos\vartheta).$$ (9.46)

Similarly, we obtain

$$g^+(\vartheta, \varphi) = \frac{1}{2iK} \sum_{l=1}^{\infty} [\exp(2i\eta_l^-) - \exp(2i\eta_l^+)] P_l^1(\cos\vartheta) \exp(i\varphi).$$ (9.47)

For electrons with spins antiparallel to the direction of incidence (spin down), we have

$$f^-(\vartheta) = f^+(\vartheta) \tag{9.48}$$

and

$$g^-(\vartheta, \varphi) = -g^+(\vartheta, \varphi) \exp(-2i\varphi), \tag{9.49}$$

where we indicate by f^- and g^- the scattering amplitudes for the case of electrons with spin down. Once defined, the functions

$$f(\vartheta) = \sum_{l=0}^{\infty} A_l P_l(\cos \vartheta), \tag{9.50}$$

$$g(\vartheta) = \sum_{l=0}^{\infty} B_l P_l^1(\cos \vartheta), \tag{9.51}$$

where

$$A_l = \frac{1}{2iK} \{(l+1) [\exp(2i\eta_l^+) - 1] + l [\exp(2i\eta_l^-) - 1]\}, \tag{9.52}$$

$$B_l = \frac{1}{2iK} [\exp(2i\eta_l^-) - \exp(2i\eta_l^+)], \tag{9.53}$$

we have

$$f^+ = f^- = f, \tag{9.54}$$

$$g^+ = g \exp(i\varphi), \tag{9.55}$$

and

$$g^- = -g \exp(-i\varphi). \tag{9.56}$$

9.2 Elastic scattering cross-section

9.2.1 Relativistic elastic scattering cross-section

Let us note first that, for electrons with spin up, we have:

$$\frac{d\sigma}{d\Omega} = \frac{|f^+|^2 + |g^+|^2}{1+0} = |f|^2 + |g|^2 \tag{9.57}$$

as, in this case, $a_1 = 1$, $a_2 = 0$ $b_1 = f^+$, and $b_2 = g^+$. The case of electrons with spin down provides an identical result since, in this case, $a_1 = 0$, $a_2 = 1$ $b_1 = g^-$, and

$b_2 = f^-$, so that

$$\frac{d\sigma}{d\Omega} = \frac{|g^-|^2 + |f^-|^2}{0 + 1} = |f|^2 + |g|^2 . \tag{9.58}$$

For an arbitrary spin direction, the electron incident plane wave will be given by $\Psi_1 = A\exp(iKz)$ and $\Psi_2 = B\exp(iKz)$. As a consequence, $a_1 = A$ and $a_2 = B$. Furthermore,

$$b_1 = Af^+ + Bg^- = Af - Bg\exp(-i\varphi), \tag{9.59}$$
$$b_2 = Ag^+ + Bf^- = Bf + Ag\exp(i\varphi). ^1 \tag{9.60}$$

Thus we have

$$|b_1|^2 = |A|^2|f|^2 + |B|^2|g|^2 - AB^*fg^*\exp(i\varphi) - A^*Bf^*g\exp(-i\varphi)$$

and

$$|b_2|^2 = |B|^2|f|^2 + |A|^2|g|^2 + AB^*f^*g\exp(i\varphi) + A^*Bfg^*\exp(-i\varphi),$$

so that

$$|b_1|^2 + |b_2|^2$$
$$= (|f|^2 + |g|^2)\left\{|A|^2 + |B|^2 + i\left(i\frac{fg^* - f^*g}{|f|^2 + |g|^2}\right)[AB^*\exp(i\varphi) - A^*B\exp(-i\varphi)]\right\}.$$

Therefore,

$$\frac{d\sigma}{d\Omega} = \frac{|b_1|^2 + |b_2|^2}{|a_1|^2 + |a_2|^2}$$
$$= (|f|^2 + |g|^2)\left\{1 + iS(\vartheta)\left[\frac{AB^*\exp(i\varphi) - A^*B\exp(-i\varphi)}{|A|^2 + |B|^2}\right]\right\}, \tag{9.61}$$

where we have introduced the Sherman function $S(\vartheta)$ defined by

$$S(\vartheta) = i\frac{fg^* - f^*g}{|f|^2 + |g|^2} . \tag{9.62}$$

If σ_1, σ_2 and σ_3 are the Pauli matrices and η is the two-component spinor

$$\eta = \begin{pmatrix} A/\sqrt{|A|^2 + |B|^2} \\ B/\sqrt{|A|^2 + |B|^2} \end{pmatrix} = \begin{pmatrix} \alpha \\ \beta \end{pmatrix}, \tag{9.63}$$

---

1 Please note that, from Eqs. (9.59) and (9.60), we obtain $b_1 = f^+$, $b_2 = g^+$ if $A = a_1 = 1$ and $B = a_2 = 0$, and $b_1 = g^-$, $b_2 = f^-$ if $A = a_1 = 0$ and $B = a_2 = 1$.

$$\eta^{\dagger} = \left(\frac{A^*}{\sqrt{|A|^2 + |B|^2}} \quad \frac{B^*}{\sqrt{|A|^2 + |B|^2}} \right) = (\alpha^*, \beta^*),$$ (9.64)

then

$$i\frac{AB^* \exp(i\varphi) - A^*B \exp(-i\varphi)}{|A|^2 + |B|^2} = \eta^{\dagger}(\sigma_y \cos \varphi - \sigma_x \sin \varphi)\eta.$$ (9.65)

Indeed,

$$\sigma_y \cos \varphi - \sigma_x \sin \varphi = \begin{pmatrix} 0 & -i \cos \varphi \\ i \cos \varphi & 0 \end{pmatrix} - \begin{pmatrix} 0 & \sin \varphi \\ \sin \varphi & 0 \end{pmatrix}$$

$$= \begin{pmatrix} 0 & -\sin \varphi - i \cos \varphi \\ -\sin \varphi + i \cos \varphi & 0 \end{pmatrix} = i \begin{pmatrix} 0 & -e^{-i\varphi} \\ e^{i\varphi} & 0 \end{pmatrix}.$$

Thus, we have

$$\eta^{\dagger}(\sigma_y \cos \varphi - \sigma_x \sin \varphi)\eta = i(\alpha^*, \beta^*) \begin{pmatrix} 0 & -e^{-i\varphi} \\ e^{i\varphi} & 0 \end{pmatrix} \begin{pmatrix} \alpha \\ \beta \end{pmatrix}$$

$$= i(\alpha\beta^* e^{i\varphi} - \alpha^*\beta e^{-i\varphi}) = i\frac{AB^* \exp(i\varphi) - A^*B \exp(-i\varphi)}{|A|^2 + |B|^2}.$$

Since the z axis has been chosen along the direction of electron incidence, the unit vector perpendicular to the plane of scattering is given by

$$\hat{\mathbf{n}} = \begin{pmatrix} -\sin \varphi \\ \cos \varphi \\ 0 \end{pmatrix},$$ (9.66)

and, hence,

$$\boldsymbol{\sigma} \cdot \hat{\mathbf{n}} = \sigma_y \cos \varphi - \sigma_x \sin \varphi.$$ (9.67)

Thus, we can write

$$\eta^{\dagger}(\sigma_y \cos \varphi - \sigma_x \sin \varphi)\eta = \mathbf{P} \cdot \hat{\mathbf{n}},$$ (9.68)

where

$$\mathbf{P} = \langle \boldsymbol{\sigma} \rangle$$ (9.69)

is the initial spin-polarization vector of the electron beam. Thus, the differential elastic scattering cross-section can also be written as

$$\frac{d\sigma}{d\Omega} = (|f|^2 + |g|^2)[1 + S(\vartheta)\mathbf{P} \cdot \hat{\mathbf{n}}].$$ (9.70)

Eq. (9.70), obtained here for pure spin states (i. e. for fully polarized beams, so that $|\mathbf{P}| = 1$), is valid also for a mixture of spin states (with any degree of polarization, i. e. $0 \le |\mathbf{P}| \le 1$). Note that, if the beam is completely unpolarized, then $P = 0$ and

$$\frac{d\sigma}{d\Omega} = |f|^2 + |g|^2. \tag{9.71}$$

The total elastic scattering cross-section σ and the transport elastic scattering cross-section σ_{tr} are defined by

$$\sigma = 2\pi \int_0^\pi \frac{d\sigma}{d\Omega} \sin\vartheta \, d\vartheta, \tag{9.72}$$

$$\sigma_{tr} = 2\pi \int_0^\pi (1 - \cos\vartheta)\frac{d\sigma}{d\Omega} \sin\vartheta \, d\vartheta \tag{9.73}$$

and can be easily calculated by numerical integration.

9.2.2 Nonrelativistic limit

By imposing

$$\eta_l^- = \eta_l^+ = \eta_l \tag{9.74}$$

in the previous equations, we obtain the nonrelativistic result. Indeed, in this case, we have

$$\begin{aligned}
\mathcal{A}_l &= \frac{1}{2iK}\{(l + 1)[\exp(2i\eta_l) - 1] + l[\exp(2i\eta_l) - 1]\} \\
&= \frac{1}{2iK}(2l + 1)[\exp(2i\eta_l) - 1], \tag{9.75} \\
\mathcal{B}_l &= 0, \tag{9.76}
\end{aligned}$$

so that

$$\begin{aligned}
f(\vartheta) &= \frac{1}{2iK}\sum_{l=0}^\infty (2l + 1)[\exp(2i\eta_l) - 1]P_l(\cos\vartheta) \\
&= \frac{1}{K}\sum_{l=0}^\infty (2l + 1)\exp(i\eta_l)\sin\eta_l P_l(\cos\vartheta), \tag{9.77} \\
g(\vartheta) &= 0, \tag{9.78}
\end{aligned}$$

and

$$\frac{d\sigma}{d\Omega} = |f|^2. \tag{9.79}$$

9.3 Phase-shift calculation

9.3.1 Lin, Sherman, and Percus transformation

Let us perform the following transformation (Lin, Sherman, and Percus [17]):

$$F_l^\pm(r) = a_l^\pm(r)\frac{\sin\phi_l^\pm(r)}{r}, \tag{9.80}$$

$$G_l^\pm(r) = a_l^\pm(r)\frac{\cos\phi_l^\pm(r)}{r}. \tag{9.81}$$

Equations (9.9) and (9.10) become

$$[E + m - V(r)]\tan\phi_l^\pm(r) + \frac{1}{a_l^\pm(r)}\frac{da_l^\pm(r)}{dr}$$

$$- \tan\phi_l^\pm(r)\frac{d\phi_l^\pm(r)}{dr} + \frac{k}{r} = 0, \tag{9.82}$$

$$-[E - m - V(r)]\cot\phi_l^\pm(r) + \frac{1}{a_l^\pm(r)}\frac{da_l^\pm(r)}{dr}$$

$$+ \cot\phi_l^\pm(r)\frac{d\phi_l^\pm(r)}{dr} - \frac{k}{r} = 0. \tag{9.83}$$

From these equations, we obtain

$$\frac{d\phi_l^\pm(r)}{dr} = E + m - V(r) + \frac{1}{a_l^\pm(r)}\frac{da_l^\pm(r)}{dr}\frac{1}{\tan\phi_l^\pm(r)} + \frac{k}{r}\frac{1}{\tan\phi_l^\pm(r)}, \tag{9.84}$$

and

$$\frac{d\phi_l^\pm(r)}{dr} = E - m - V(r) - \frac{1}{a_l^\pm(r)}\frac{da_l^\pm(r)}{dr}\frac{1}{\cot\phi_l^\pm(r)} + \frac{k}{r}\frac{1}{\cot\phi_l^\pm(r)}. \tag{9.85}$$

As a consequence,

$$2m + \frac{1}{a_l^\pm(r)}\frac{da_l^\pm(r)}{dr}\left[\frac{1}{\tan\phi_l^\pm(r)} + \frac{1}{\cot\phi_l^\pm(r)}\right] + \frac{k}{r}\left[\frac{1}{\tan\phi_l^\pm(r)} - \frac{1}{\cot\phi_l^\pm(r)}\right]$$

$$= 2m + \frac{1}{a_l^\pm(r)}\frac{da_l^\pm(r)}{dr}\frac{2}{\sin 2\phi_l^\pm(r)} + \frac{k}{r}\frac{2\cos 2\phi_l^\pm(r)}{\sin 2\phi_l^\pm(r)} = 0,$$

so that

$$\frac{1}{a_l^\pm(r)}\frac{da_l^\pm(r)}{dr} = -m\sin 2\phi_l^\pm(r) - \frac{k}{r}\cos 2\phi_l^\pm(r). \tag{9.86}$$

Combining now Eqs. (9.84) and (9.86), we obtain

$$\frac{d\phi_l^{\pm}(r)}{dr} = E + m - V(r) + \left[-m \sin 2\phi_l^{\pm}(r) - \frac{k}{r} \cos 2\phi_l^{\pm}(r)\right] \cot \phi_l^{\pm}(r) + \frac{k}{r} \cot \phi_l^{\pm}(r)$$

$$= E + m \left[1 - \sin 2\phi_l^{\pm}(r) \cot \phi_l^{\pm}(r)\right] + \frac{k}{r} \cot \phi_l^{\pm}(r) \left[1 - \cos 2\phi_l^{\pm}(r)\right] - V(r).$$

Since

$$1 - \sin 2\phi_l^{\pm}(r) \cot \phi_l^{\pm}(r) = -\cos 2\phi_l^{\pm}(r)$$

and

$$\cot \phi_l^{\pm}(r) \left[1 - \cos 2\phi_l^{\pm}(r)\right] = \sin 2\phi_l^{\pm}(r),$$

we conclude that

$$\frac{d\phi_l^{\pm}(r)}{dr} = \frac{k}{r} \sin 2\phi_l^{\pm}(r) - m \cos 2\phi_l^{\pm}(r) + E - V(r). \qquad (9.87)$$

9.3.2 Phase shifts

Let us now calculate the phase shifts. Examining Eq. (9.81), we obtain

$$G_l'^{\pm} = \frac{a_l'^{\pm} \cos \phi_l^{\pm}(r)}{r} - \frac{a_l^{\pm}}{r} \sin \phi_l^{\pm}(r) \phi_l'^{\pm}(r) - \frac{a_l^{\pm} \cos \phi_l^{\pm}(r)}{r^2}, \qquad (9.88)$$

so that

$$\frac{G_l'^{\pm}}{G_l^{\pm}} = \frac{a_l'^{\pm}}{a_l^{\pm}} - \phi_l'^{\pm}(r) \tan \phi_l^{\pm}(r) - \frac{1}{r}. \qquad (9.89)$$

Thus,

$$\frac{G_l'^{\pm}}{G_l^{\pm}} = -m \sin 2\phi_l^{\pm}(r) - \frac{k}{r} \cos 2\phi_l^{\pm}(r)$$

$$- \left[\frac{k}{r} \sin 2\phi_l^{\pm}(r) - m \cos 2\phi_l^{\pm}(r) + E - V(r)\right] \tan \phi_l^{\pm}(r) - \frac{1}{r}$$

$$= -\frac{k}{r} \left[\cos 2\phi_l^{\pm}(r) + \sin 2\phi_l^{\pm}(r) \tan \phi_l^{\pm}(r)\right]$$

$$- m \left[\sin 2\phi_l^{\pm}(r) - \cos 2\phi_l^{\pm}(r) \tan \phi_l^{\pm}(r)\right] - [E - V(r)] \tan \phi_l^{\pm}(r) - \frac{1}{r}.$$

Since

$$\cos 2\phi_l^\pm(r) + \sin 2\phi_l^\pm(r) \tan \phi_l^\pm(r)$$
$$= \cos^2 \phi_l^\pm(r) - \sin^2 \phi_l^\pm(r) + 2\sin^2 \phi_l^\pm(r) = 1$$

and

$$\sin 2\phi_l^\pm(r) - \cos 2\phi_l^\pm(r) \tan \phi_l^\pm(r)$$
$$= 2\sin \phi_l^\pm(r) \cos \phi_l^\pm(r) - \sin \phi_l^\pm(r) \cos \phi_l^\pm(r) + \tan \phi_l^\pm(r) \sin^2 \phi_l^\pm(r)$$
$$= \sin \phi_l^\pm(r) \left[\frac{\cos^2 \phi_l^\pm(r) + \sin^2 \phi_l^\pm(r)}{\cos \phi_l^\pm(r)} \right] = \tan \phi_l^\pm(r),$$

we have

$$\frac{G_l'^\pm}{G_l^\pm} = -(E + m - V) \tan \phi_l^\pm(r) - \frac{1 + k}{r}. \tag{9.90}$$

Keep in mind that the asymptotic form of the solution in the regions corresponding to large values of r for which $V(r) \approx 0$ is given by Eq. (9.24), repeated here for the reader's convenience:

$$G_l^\pm \underset{r \to \infty}{\sim} j_l(Kr) \cos \eta_l^\pm - n_l(Kr) \sin \eta_l^\pm,$$

where $K^2 = E^2 - m^2$, $\eta_l^\pm$ is the lth phase shift, and j_l and n_l are, respectively, the regular and irregular spherical Bessel functions. Therefore,

$$\frac{G_l'^\pm}{G_l^\pm} = \frac{Kj_l'(Kr) \cos \eta_l^\pm - Kn_l'(Kr) \sin \eta_l^\pm}{j_l(Kr) \cos \eta_l^\pm - n_l(Kr) \sin \eta_l^\pm}. \tag{9.91}$$

Taking into account the properties of the Bessel functions, Eqs. (2.45) and (2.46),

$$xj_{l-1} - (2l + 1)j_l + xj_{l+1} = 0, \tag{9.92}$$

$$xj_{l-1} - (l + 1)j_l - x\frac{dj_l}{dx} = 0, \tag{9.93}$$

$$xn_{l-1} - (2l + 1)n_l + xn_{l+1} = 0, \tag{9.94}$$

$$xn_{l-1} - (l + 1)n_l - x\frac{dn_l}{dx} = 0, \tag{9.95}$$

it is easy to see that

$$j_l'(x) = \frac{l}{x}j_l(x) - j_{l+1}(x), \tag{9.96}$$

$$n_l'(x) = \frac{l}{x}n_l(x) - n_{l+1}(x). \tag{9.97}$$

As a consequence, we have

$$
\frac{G_l'^{\pm}}{G_l^{\pm}} = \frac{K[\frac{l}{Kr}j_l(Kr) - j_{l+1}(Kr)]\cos\eta_l^{\pm} - K[\frac{l}{Kr}n_l(Kr) - n_{l+1}(Kr)]\sin\eta_l^{\pm}}{j_l(Kr)\cos\eta_l^{\pm} - n_l(Kr)\sin\eta_l^{\pm}}
$$

$$
= \frac{(l/r)j_l(Kr) - Kj_{l+1}(Kr) - [(l/r)n_l(Kr) - Kn_{l+1}(Kr)]\tan\eta_l^{\pm}}{j_l(Kr) - n_l(Kr)\tan\eta_l^{\pm}},
$$

so that

$$
\tan\eta_l^{\pm} = \frac{(l/r)j_l(Kr) - Kj_{l+1}(Kr) - j_l(Kr)(G_l'^{\pm}/G_l^{\pm})}{(l/r)n_l(Kr) - Kn_{l+1}(Kr) - n_l(Kr)(G_l'^{\pm}/G_l^{\pm})}. \tag{9.98}
$$

Let us now define

$$
\tilde{\phi}_l^{\pm} = \lim_{r\to\infty} \phi_l^{\pm}(r). \tag{9.99}
$$

For large values of r, $V(r) \approx 0$ and Eq. (9.90) becomes

$$
\frac{G_l'^{\pm}}{G_l^{\pm}} = -(E + m)\tan\tilde{\phi}_l^{\pm} - \frac{1+k}{r}, \tag{9.100}
$$

so that

$$
(l/r)j_l(Kr) - Kj_{l+1}(Kr) - j_l(Kr)(G_l'^{\pm}/G_l^{\pm})
$$
$$
= -Kj_{l+1}(Kr) + j_l(Kr)[(E + m)\tan\tilde{\phi}_l^{\pm} + (1 + l + k)/r],
$$
$$
(l/r)n_l(Kr) - Kn_{l+1}(Kr) - n_l(Kr)(G_l'^{\pm}/G_l^{\pm})
$$
$$
= -Kn_{l+1}(Kr) + n_l(Kr)[(E + m)\tan\tilde{\phi}_l^{\pm} + (1 + l + k)/r].
$$

Therefore,

$$
\tan\eta_l^{\pm} = \frac{Kj_{l+1}(Kr) - j_l(Kr)[(E + m)\tan\tilde{\phi}_l^{\pm} + (1 + l + k)/r]}{Kn_{l+1}(Kr) - n_l(Kr)[(E + m)\tan\tilde{\phi}_l^{\pm} + (1 + l + k)/r]}. \tag{9.101}
$$

Using the previous equation, we can calculate the phase shifts of the scattered wave and, therefore, the differential elastic scattering cross-section.

9.3.3 Numerical approach

To calculate phase shifts using Eq. (9.101), it is necessary to know $\tilde{\phi}_l^{\pm}$. This means that we need to numerically solve Eq. (9.87) up to a radius r_{max}, where the potential can be considered negligible. The numerical integration of Eq. (9.87) using, for example, the fourth-order Runge–Kutta method, requires knowing the function $\phi_l^{\pm}$ when r is very small. It is also necessary to know the value of the potential for very small values of r. For $0 < r < \hbar/mc$, the spherical symmetric electrostatic potential experienced by an electron at distance r from the nucleus, $V(r)$, may be approximated by the following equation:

$$V(r) \underset{r \to 0}{\sim} -\frac{Z_0 + Z_1 r + Z_2 r^2 + Z_3 r^3}{r}. \tag{9.102}$$

Let us express, as usual, the electrostatic potential as the product of the potential of a bare nucleus multiplied by a screening function $\xi(r)$ having the analytical form:

$$\xi(r) = \sum_{i=1}^{p} \gamma_i \exp(-\lambda_i r),$$

$$\sum_{i=1}^{p} \gamma_i = 1.$$

Using this equation, we can easily evaluate Z_0, Z_1, Z_2, and Z_3 as:

$$Z_0 = Ze^2 \sum_i \gamma_i = Ze^2, \tag{9.103}$$

$$Z_1 = -Z_0 \sum_{i=1}^{p} \lambda_i \gamma_i, \tag{9.104}$$

$$Z_2 = \frac{Z_0}{2} \sum_{i=1}^{p} \lambda_i^2 \gamma_i, \tag{9.105}$$

$$Z_3 = -\frac{Z_0}{6} \sum_{i=1}^{p} \lambda_i^3 \gamma_i. \tag{9.106}$$

Expanding $\phi_l^{\pm}$ as a power series, we obtain

$$\phi_l^{\pm}(r) = \phi_{l0}^{\pm} + \phi_{l1}^{\pm} r + \phi_{l2}^{\pm} r^2 + \phi_{l3}^{\pm} r^3 + \cdots. \tag{9.107}$$

The relationships between the coefficients of this expansion and Z_0, Z_1, Z_2, Z_3 are the following:[2]

$$\sin 2\phi_{l0}^{\pm} = -\frac{Z_0}{k}, \tag{9.108}$$

$$\phi_{l1}^{\pm} = \frac{E + Z_1 - m \cos 2\phi_{l0}^{\pm}}{1 - 2k \cos 2\phi_{l0}^{\pm}}, \tag{9.109}$$

---

2 The simple but rather lengthy demonstration, which can be obtained by substituting Eqs. (9.102) and (9.107) in Eq. (9.87), is left to the reader as a useful exercise.

$$\phi_{l2}^{\pm} = \frac{2\phi_{l1}^{\pm}\sin 2\phi_{l0}^{\pm}(m - k\phi_{l1}^{\pm}) + Z_2}{2 - 2k\cos 2\phi_{l0}^{\pm}}, \tag{9.110}$$

$$\phi_{l3}^{\pm} = \frac{2\phi_{l2}^{\pm}\sin 2\phi_{l0}^{\pm}(m - 2k\phi_{l1}^{\pm}) + 2\phi_{l1}^{\pm 2}\cos 2\phi_{l0}^{\pm}[m - (2/3)k\phi_{l1}^{\pm}] + Z_3}{3 - 2k\cos 2\phi_{l0}^{\pm}}, \tag{9.111}$$

with the extra conditions:

$$0 \le 2\phi_{l0}^{\pm} \le \frac{1}{2}\pi \tag{9.112}$$

if $k < 0$, and

$$\pi \le 2\phi_{l0}^{\pm} \le \frac{3}{2}\pi \tag{9.113}$$

if $k > 0$.

If we use the mass of the electron as unit of energy (and mass), then $m = 1$ and[3]

$$W = \frac{E}{m}. \tag{9.114}$$

Equations (9.109), (9.110), and (9.111) become

$$\phi_{l1}^{\pm} = \frac{W + Z_1 - \cos 2\phi_{l0}^{\pm}}{1 - 2k\cos 2\phi_{l0}^{\pm}}, \tag{9.115}$$

$$\phi_{l2}^{\pm} = \frac{2\phi_{l1}^{\pm}\sin 2\phi_{l0}^{\pm}(1 - k\phi_{l1}^{\pm}) + Z_2}{2 - 2k\cos 2\phi_{l0}^{\pm}}, \tag{9.116}$$

$$\phi_{l3}^{\pm} = \frac{2\phi_{l2}^{\pm}\sin 2\phi_{l0}^{\pm}(1 - 2k\phi_{l1}^{\pm}) + 2\phi_{l1}^{\pm 2}\cos 2\phi_{l0}^{\pm}[1 - (2/3)k\phi_{l1}^{\pm}] + Z_3}{3 - 2k\cos 2\phi_{l0}^{\pm}}. \tag{9.117}$$

Please note that, with this choice of energy units, the equation for calculating the phase shifts has the form:

$$\tan \eta_l^{\pm} = \frac{Kj_{l+1}(Kr) - j_l(Kr)[(W + 1)\tan \phi_l^{\pm} + (1 + l + k)/r]}{Kn_{l+1}(Kr) - n_l(Kr)[(W + 1)\tan \phi_l^{\pm} + (1 + l + k)/r]}. \tag{9.118}$$

---

3 We use here the symbol W for indicating the energy to help keep in mind that it is expressed in m units.

9.4 Electron–molecule elastic scattering

According to Salvat et al. [23], when the target is a molecule, instead of an atom, and the spin-polarization is null, then

$$\frac{d\sigma}{d\Omega} = \sum_{m,n} \exp(i\mathbf{q} \cdot \mathbf{r}_{mn}) \left[f_m(\vartheta)f_n^*(\vartheta) + g_m(\vartheta)g_n^*(\vartheta) \right], \qquad (9.119)$$

where $\mathbf{q}$ is the momentum transfer, $\mathbf{r}_{mn} = \mathbf{r}_m - \mathbf{r}_n$, and $\mathbf{r}_m$ is the position vector of the mth atom in the molecule. Please note that this equation, in case $m = n$, so that

$$\mathbf{r}_{mn} = \mathbf{r}_m - \mathbf{r}_m = 0,$$

reduces to the previous one, describing electron–atom elastic scattering:

$$\frac{d\sigma}{d\Omega} = \exp(0) \left[f_m(\vartheta)f_m^*(\vartheta) + g_m(\vartheta)g_m^*(\vartheta) \right] = |f_m(\vartheta)|^2 + |g_m(\vartheta)|^2. \qquad (9.120)$$

If the molecules in the target are randomly oriented, we are allowed to average out all the orientations. Please note that, if we indicate with α the angle between $\mathbf{q}$ and $\mathbf{r}_{mn}$, we have

$$\frac{\int_0^\pi 2\pi \sin\alpha \, e^{iqr_{mn}\cos\alpha} \, d\alpha}{\int_0^\pi 2\pi \sin\alpha \, d\alpha} = \frac{\int_{-1}^1 e^{iqr_{mn}u} \, du}{\int_{-1}^1 du}$$

$$= \frac{1}{2} \int_{-1}^1 \cos(q\, r_{mn}\, u) \, du = \frac{\sin qr_{mn}}{qr_{mn}},$$

so that Eq. (9.119) can be simplified as follows:

$$\frac{d\sigma}{d\Omega} = \sum_{m,n} \frac{\sin qr_{mn}}{qr_{mn}} \left[f_m(\vartheta)f_n^*(\vartheta) + g_m(\vartheta)g_n^*(\vartheta) \right], \qquad (9.121)$$

where the modulus of the momentum transfer is given by

$$q = 2K \sin(\vartheta/2) \qquad (9.122)$$

and K is the modulus of the momentum of the projectile. Let us consider now a randomly oriented diatomic molecule. In this case we have

$$\left(\frac{d\sigma}{d\Omega} \right) = \frac{\sin qr_{11}}{qr_{11}}(f_1f_1^* + g_1g_1^*) + \frac{\sin qr_{12}}{qr_{12}}(f_1f_2^* + g_1g_2^*)$$

$$+ \frac{\sin qr_{21}}{qr_{21}}(f_2f_1^* + g_2g_1^*) + \frac{\sin qr_{22}}{qr_{22}}(f_2f_2^* + g_2g_2^*). \qquad (9.123)$$

Because

$$\frac{\sin qr_{11}}{qr_{11}} = \frac{\sin qr_{22}}{qr_{22}} = \lim_{x \to 0} \frac{\sin x}{x} = 1, \tag{9.124}$$

and

$$f_1 f_1^* + g_1 g_1^* = |f_1|^2 + |g_1|^2 = \left(\frac{d\sigma}{d\Omega}\right)_1, \tag{9.125}$$

$$f_2 f_2^* + g_2 g_2^* = |f_2|^2 + |g_2|^2 = \left(\frac{d\sigma}{d\Omega}\right)_2, \tag{9.126}$$

we have

$$\left(\frac{d\sigma}{d\Omega}\right) = \left(\frac{d\sigma}{d\Omega}\right)_1 + \left(\frac{d\sigma}{d\Omega}\right)_2$$
$$+ \frac{\sin qr_{12}}{qr_{12}}(f_1 f_2^* + g_1 g_2^*) + \frac{\sin qr_{21}}{qr_{21}}(f_2 f_1^* + g_2 g_1^*). \tag{9.127}$$

Since

$$r_{12} = r_{21}, \tag{9.128}$$

we can also write

$$\left(\frac{d\sigma}{d\Omega}\right) = \left(\frac{d\sigma}{d\Omega}\right)_1 + \left(\frac{d\sigma}{d\Omega}\right)_2$$
$$+ \frac{\sin qr_{12}}{qr_{12}}(f_1 f_2^* + f_2 f_1^* + g_1 g_2^* + g_2 g_1^*). \tag{9.129}$$

Note that the frequently used *additivity approximation*:

$$\left(\frac{d\sigma}{d\Omega}\right) = \left(\frac{d\sigma}{d\Omega}\right)_1 + \left(\frac{d\sigma}{d\Omega}\right)_2, \tag{9.130}$$

neglects the term $(\sin qr_{12}/qr_{12})(f_1 f_2^* + f_2 f_1^* + g_1 g_2^* + g_2 g_1^*)$. In the case of a three-atomic molecule, we similarly obtain

$$\left(\frac{d\sigma}{d\Omega}\right) = \left(\frac{d\sigma}{d\Omega}\right)_1 + \left(\frac{d\sigma}{d\Omega}\right)_2 + \left(\frac{d\sigma}{d\Omega}\right)_3$$
$$+ \frac{\sin qr_{12}}{qr_{12}}(f_1 f_2^* + f_2 f_1^* + g_1 g_2^* + g_2 g_1^*)$$
$$+ \frac{\sin qr_{23}}{qr_{23}}(f_2 f_3^* + f_3 f_2^* + g_2 g_3^* + g_3 g_2^*)$$
$$+ \frac{\sin qr_{31}}{qr_{31}}(f_3 f_1^* + f_1 f_3^* + g_3 g_1^* + g_1 g_3^*). \tag{9.131}$$

10 Density matrix and spin-polarization phenomena

In this chapter, we will deal with the problem of the spin-polarization of electron beams and show that it can be adequately and satisfactorily addressed only in the setting of a quantum-relativistic theory such as the Dirac theory [5, 14].

10.1 The density matrix

We will first introduce a very important concept described by the so-called *density matrix*. The spin orientation of an electron beam is described by a probability distribution instead of by a single vector of state. Actually, an electron beam is a quantum system whose state is a superposition of number of N substates. Every substate is a *pure* state of spin. The overall state is heterogeneous or *mixed*. We will use the Dirac notation and indicate with $|a\rangle$ the state vectors representing the pure states composing the quantum system.[1] Let us indicate with $\{|n\rangle\}$ a complete set of orthonormal eigenvectors.[2] We can hence expand each pure state $|a\rangle$ using the complete set of orthonormal eigenvectors $|n\rangle$:

$$|a\rangle = \sum_n c_n |n\rangle,$$ (10.1)

where

$$c_n = \langle n|a\rangle$$ (10.2)

are the projections of the pure state $|a\rangle$ onto the orthonormal eigenvectors $|n\rangle$.[3] The expectation value of any operator A is given by

$$\langle A\rangle = \sum_{a=1}^{N} p_a \langle a|A|a\rangle,$$ (10.3)

---

1 Pure states $|a\rangle$ are not, in general, orthogonal to each other. Anyway, they are assumed to be normalized vectors.

2 Let us remind our readers that a complete set of basis states $\{|n\rangle\}$ is a set of orthonormal eigenvectors of some complete set of mutually commuting operators. The fact that the eigenvectors are orthonormal implies that $\langle n|n'\rangle = \delta_{nn'}$. The set of basis states is complete, so that $\sum_n |n\rangle\langle n| = I$, where we have indicated with I the identity operator.

3 Note that $\sum_n |c_n|^2 = 1$, due to the completeness of the basis states and to the normalization of the pure states. In fact $\sum_n |c_n|^2 = \sum_n \langle a|n\rangle\langle n|a\rangle = \langle a|(\sum_n |n\rangle\langle n|)|a\rangle = \langle a|I|a\rangle = \langle a|a\rangle = 1$.

https://doi.org/10.1515/9783110675375-010

where p_a is the probability of obtaining the pure state $|a\rangle$.[4] The mean value of the operator A in the pure state $|a\rangle$ is given by

$$\langle A\rangle_a = \langle a|A|a\rangle, \tag{10.4}$$

so that we can write

$$\langle A\rangle = \sum_{a=1}^{N} p_a\langle A\rangle_a. \tag{10.5}$$

The statistical weights p_a of the pure states $|a\rangle$ allow us to introduce the density matrix ρ, defined as

$$\rho = \sum_{a=1}^{N} |a\rangle p_a\langle a|. \tag{10.6}$$

As we can see from the definition of ρ, the density matrix is Hermitian, i. e., $\rho^\dagger = \rho$: This means that, using a unitary transformation, the density matrix can always be diagonalized. We can easily calculate the matrix elements of ρ:

$$\rho_{nm} = \langle n|\rho|m\rangle = \sum_{a=1}^{N} \langle n|a\rangle p_a\langle a|m\rangle = \sum_{a=1}^{N} p_a c_n c_m^*. \tag{10.7}$$

Let us indicate with the symbol Tr the trace of a matrix. By definition, $\mathrm{Tr}(M)$ is the sum of the diagonal elements of the matrix M:

$$\mathrm{Tr}(M) = \sum_n \langle n|M|n\rangle. \tag{10.8}$$

Let us now calculate the trace of the matrix ρA. Since the set of basis states is complete,

$$\sum_m |m\rangle\langle m| = I. \tag{10.9}$$

Here we have indicated with I the identity operator. We have

$$\mathrm{Tr}(\rho A) = \sum_n \langle n|\rho A|n\rangle = \sum_n \langle n|\rho I A|n\rangle = \sum_n \sum_m \langle n|\rho|m\rangle\langle m|A|n\rangle. \tag{10.10}$$

Since

$$|a\rangle = \sum_n \langle n|a\rangle|n\rangle, \tag{10.11}$$

---

4 The probability p_a of obtaining the subsystem a satisfies the following conditions: $0 \le p_a \le 1$ and $\sum_{a=1}^{N} p_a = 1$.

thus

$$\langle A \rangle = \sum_{a=1}^{N} p_a \sum_n \sum_m \langle n|a \rangle \langle a|m \rangle \langle m|A|n \rangle . \tag{10.12}$$

To conclude, since

$$\text{Tr}(\rho A) = \sum_n \sum_m \langle n|\rho|m \rangle \langle m|A|n \rangle = \sum_{a=1}^{N} \sum_n \sum_m \langle n|a \rangle p_a \langle a|m \rangle \langle m|A|n \rangle , \tag{10.13}$$

we have

$$\text{Tr}(\rho A) = \langle A \rangle . \tag{10.14}$$

So, if we know the density matrix, we can easily obtain the mean value of any observable A. In particular, if the observable is the identity operator I, since

$$\langle I \rangle = \sum_{a=1}^{N} p_a \langle a|I|a \rangle = \sum_{a=1}^{N} p_a \langle a|a \rangle = \sum_{a=1}^{N} p_a = 1 , \tag{10.15}$$

we obtain

$$\text{Tr}(\rho) = 1 . \tag{10.16}$$

10.2 The spin-polarization

Let us now consider the spin space. As we know, the Pauli matrices together with the 2×2 identity matrix constitute a complete set of 2×2 Hermitian matrices. Therefore, the density matrix can be developed on the Pauli matrices and on the identity matrix as follows:

$$\rho = a_0 I + a_1 \sigma_x + a_2 \sigma_y + a_3 \sigma_z . \tag{10.17}$$

In this equation, the coefficients a_0, a_1, a_2, a_3 are real numbers, I is the 2×2 identity matrix, and $\sigma_x = \sigma_1, \sigma_y = \sigma_2$, and $\sigma_z = \sigma_3$ are the Pauli matrices. Let us now calculate the average values of the Pauli matrices:

$$\langle \sigma_j \rangle = \text{Tr}(\rho \sigma_j) , \tag{10.18}$$

where $j = 1, 2, 3$. Since

$$\text{Tr}(\rho) = \text{Tr}(\rho I) = \langle I \rangle = 1 , \tag{10.19}$$

and

$$\mathrm{Tr}(\sigma_j) = 0, \tag{10.20}$$

for each $j = 1, 2, 3$, we can conclude that

$$\mathrm{Tr}(\rho) = a_0 \, \mathrm{Tr}(I) = 2a_0 = 1. \tag{10.21}$$

As a consequence,

$$a_0 = \frac{1}{2}. \tag{10.22}$$

We then observe that, since

$$\sigma_x \sigma_y = i\sigma_z = i \begin{pmatrix} 1 & 0 \\ 0 & -1 \end{pmatrix}, \tag{10.23}$$

we have

$$\mathrm{Tr}(\sigma_x \sigma_y) = 0. \tag{10.24}$$

Similarly, we can see that

$$\mathrm{Tr}(\sigma_y \sigma_z) = 0 \tag{10.25}$$

and

$$\mathrm{Tr}(\sigma_z \sigma_x) = 0. \tag{10.26}$$

Thus we have, for each $j, k = 1, 2, 3$ and $j \neq k$,

$$\mathrm{Tr}(\sigma_j \sigma_k) = 0. \tag{10.27}$$

Furthermore, from

$$\sigma_j^2 = I, \tag{10.28}$$

for every $j = 1, 2, 3$, it follows that

$$\mathrm{Tr}(\sigma_j \sigma_j) = 2. \tag{10.29}$$

This means that

$$\mathrm{Tr}(\sigma_j \sigma_k) = 2\delta_{jk}, \tag{10.30}$$

and so

$$\langle \sigma_j \rangle = \text{Tr}(\rho \sigma_j) = \text{Tr}\left[\left(a_0 I + \sum_k a_k \sigma_k\right)\sigma_j\right]$$

$$= \text{Tr}\left[\sum_k a_k \sigma_k \sigma_j\right] = 2\sum_k a_k \delta_{kj} = 2 a_j. \tag{10.31}$$

Thus

$$a_j = \frac{1}{2}\langle \sigma_j \rangle. \tag{10.32}$$

The components of the spin-polarization vector are given by

$$P_j = \langle \sigma_j \rangle = \text{Tr}(\rho \, \sigma_j). \tag{10.33}$$

Using this definition, the density matrix can be expressed as

$$\rho = \frac{1}{2}\left(I + \sum_j \sigma_j P_j\right) = \frac{1}{2}(I + \boldsymbol{\sigma} \cdot \mathbf{P}), \tag{10.34}$$

or

$$\rho = \frac{1}{2}\begin{pmatrix} 1+P_3 & P_1-iP_2 \\ P_1+iP_2 & 1-P_3 \end{pmatrix}. \tag{10.35}$$

Choosing the z axis in the direction of the polarization $\mathbf{P}$, we obtain

$$P_1 = P_2 = 0, \tag{10.36}$$
$$P_3 = |\mathbf{P}| = P \tag{10.37}$$

and

$$\rho = \frac{1}{2}\begin{pmatrix} 1+P & 0 \\ 0 & 1-P \end{pmatrix}. \tag{10.38}$$

The probability that an electron is in the pure state $|a\rangle$ is p_a. The probability that the pure state $|a\rangle$ is in the state $|j\rangle$ is $|c_j|^2$. Since the diagonal element ρ_{jj} of the density matrix is given by

$$\rho_{jj} = \langle j|\rho|j\rangle = \sum_{a=1}^{N} p_a |c_j|^2, \tag{10.39}$$

it represents the probability that an electron be in the state j. Let u be the number of electrons of the beam with spin up and d the number of electrons of the beam with spin down. Thus,

$$\rho_{11} = \frac{1+P}{2} = \frac{u}{u+d} \tag{10.40}$$

is the probability that an electron is in the state 1 (spin up) and

$$\rho_{22} = \frac{1-P}{2} = \frac{d}{u+d} \tag{10.41}$$

is the probability that an electron is in the state 2 (spin down). As a consequence, we can easily express P as

$$P = \frac{u-d}{u+d}. \tag{10.42}$$

The density matrix ρ can be written as

$$\rho = (1-P)\rho_{1/2} + P\rho_1 = (1-P)\begin{pmatrix} 1/2 & 0 \\ 0 & 1/2 \end{pmatrix} + P\begin{pmatrix} 1 & 0 \\ 0 & 0 \end{pmatrix}. \tag{10.43}$$

The matrix:

$$\rho_{1/2} = \begin{pmatrix} 1/2 & 0 \\ 0 & 1/2 \end{pmatrix}, \tag{10.44}$$

describes a completely unpolarized system as the spin-up and spin-down probabilities are identical. The other matrix:

$$\rho_1 = \begin{pmatrix} 1 & 0 \\ 0 & 0 \end{pmatrix}, \tag{10.45}$$

represents a pure state with all beam electrons having up orientation of the spins and describes a beam completely polarized.

This also means that, if P is zero, the beam is completely unpolarized, while, if P is 1, the beam is completely polarized. All the other situations, corresponding to $0 < P < 1$, represent partially polarized beams.

10.3 Polarization change following a collision

We remind our readers that the asymptotic form of the scattered wave is given by the superposition of a plane wave and a spherical wave:

$$\Psi_j \underset{r\to\infty}{\sim} a_j \exp(iKz) + b_j(\vartheta,\varphi)\frac{\exp(iKr)}{r},$$

and the differential elastic scattering cross section is given by

$$\frac{d\sigma}{d\Omega} = \frac{|b_1|^2 + |b_2|^2}{|a_1|^2 + |a_2|^2}.$$

When the spin is parallel to the incidence direction (spin up), $a_1 = 1$, $a_2 = 0$, $b_1 = f^+(\vartheta, \varphi)$, $b_2 = g^+(\vartheta, \varphi)$. When the spin is antiparallel to the incidence direction (spin down), then $a_1 = 0$, $a_2 = 1$, $b_1 = g^-(\vartheta, \varphi)$, $b_2 = f^-(\vartheta, \varphi)$. For an arbitrary direction of spin, $a_1 = A$, $a_2 = B$, $b_1 = Af^+ + Bg^-$, $b_2 = Ag^+ + Bf^-$. We have observed that the relationships between the functions $f^+(\vartheta, \varphi)$, $g^+(\vartheta, \varphi)$, $f^-(\vartheta, \varphi)$, $g^-(\vartheta, \varphi)$ and the functions $f(\vartheta)$ and $g(\vartheta)$ are given by

$$f^+(\vartheta, \varphi) = f^+(\vartheta) = f(\vartheta),$$
$$g^+(\vartheta, \varphi) = g(\vartheta) \exp(i\varphi),$$
$$g^-(\vartheta, \varphi) = -g(\vartheta) \exp(-i\varphi).$$

As a consequence, we have, for an arbitrary direction of spin,

$$b_1(\vartheta, \varphi) = Af(\vartheta) - Bg(\vartheta) \exp(-i\varphi),$$
$$b_2(\vartheta, \varphi) = Bf(\vartheta) + Ag(\vartheta) \exp(i\varphi).$$

It is convenient to introduce the matrix $M(\vartheta, \varphi)$:

$$M(\vartheta, \varphi) = \begin{pmatrix} f(\vartheta) & -g(\vartheta) \exp(-i\varphi) \\ g(\vartheta) \exp(i\varphi) & f(\vartheta) \end{pmatrix}. \tag{10.46}$$

If we indicate the initial state as

$$|\eta\rangle = \begin{pmatrix} a_1 \\ a_2 \end{pmatrix} = \begin{pmatrix} A \\ B \end{pmatrix} \tag{10.47}$$

and the final state as

$$|\eta_f\rangle = q \begin{pmatrix} b_1(\vartheta, \varphi) \\ b_2(\vartheta, \varphi) \end{pmatrix} = q \begin{pmatrix} Af(\vartheta) - Bg(\vartheta) \exp(-i\varphi) \\ Ag(\vartheta) \exp(i\varphi) + Bf(\vartheta) \end{pmatrix}, \tag{10.48}$$

where q is a normalization constant, we can thus write

$$q M |\eta\rangle = q \begin{pmatrix} f(\vartheta) & -g(\vartheta) \exp(-i\varphi) \\ g(\vartheta) \exp(i\varphi) & f(\vartheta) \end{pmatrix} \begin{pmatrix} A \\ B \end{pmatrix}$$
$$= q \begin{pmatrix} Af(\vartheta) - Bg(\vartheta) \exp(-i\varphi) \\ Ag(\vartheta) \exp(i\varphi) + Bf(\vartheta) \end{pmatrix}, \tag{10.49}$$

so that

$$|\eta_f\rangle = qM\,|\eta\rangle.$$ (10.50)

To calculate the normalization constant q, we observe that the final density matrix ρ_f and the initial density matrix ρ are related by the relationship:

$$\rho_f = |q|^2\, M\rho M^\dagger.$$ (10.51)

Indeed,

$$\begin{aligned}
\rho_f &= \sum_{a=\pm 1/2} |\eta_f^a\rangle\, p_a \langle \xi_f^a| \\
&= |q|^2 \sum_{a=\pm 1/2} M|\eta^a\rangle\, p_a \langle \eta^a| M^\dagger \\
&= |q|^2\, M\left(\sum_{a=\pm 1/2} |\eta^a\rangle\, p_a \langle \eta^a| \right) M^\dagger \\
&= |q|^2\, M\rho M^\dagger,
\end{aligned}$$ (10.52)

where $|\eta^a\rangle$ is the initial state and $|\eta_f^a\rangle$ is the final state ($a = -1/2, +1/2$). Since

$$1 = \mathrm{Tr}(\rho_f) = |q|^2\, \mathrm{Tr}(M\rho M^\dagger),$$ (10.53)

the squared modulus of the normalization constant q is equal to

$$|q|^2 = \frac{1}{\mathrm{Tr}(M\rho M^\dagger)},$$ (10.54)

so that

$$\rho_f = \frac{M\rho M^\dagger}{\mathrm{Tr}(M\rho M^\dagger)}.$$ (10.55)

We can now calculate the spin-polarization vector $\mathbf{P}_f$ after scattering. If $\mathbf{P}$ is the spin-polarization vector before scattering, then

$$\mathbf{P}_f = \mathrm{Tr}(\rho_f\,\boldsymbol{\sigma}) = \frac{\mathrm{Tr}(M\,M^\dagger\,\boldsymbol{\sigma}) + \mathrm{Tr}(M\,(\boldsymbol{\sigma}\cdot\mathbf{P})\,M^\dagger\,\boldsymbol{\sigma})}{\mathrm{Tr}(M\,M^\dagger) + \mathrm{Tr}(M\,(\boldsymbol{\sigma}\cdot\mathbf{P})\,M^\dagger)},$$ (10.56)

where we used Eq. (10.34).

10.4 Polarization of an electron beam initially not polarized

If the beam is initially not polarized, Eq. (10.56) becomes

$$\mathbf{P}_f = \frac{\mathrm{Tr}(M\,M^\dagger\,\boldsymbol{\sigma})}{\mathrm{Tr}(M\,M^\dagger)}.$$

(10.57)

From this equation, we see that the process of scattering polarizes a beam initially not polarized. We will prove that, in this case, the modulus of the final polarization is equal to the Sherman function and that its direction is perpendicular to the scattering plane. Let us remind our readers, first of all, that

$$\boldsymbol{\sigma}\cdot\hat{\mathbf{n}} = i\begin{pmatrix} 0 & -\exp(-i\varphi) \\ \exp(i\varphi) & 0 \end{pmatrix},$$

(10.58)

where we have indicated with $\hat{\mathbf{n}}$ the unit vector normal to the scattering plane:

$$\hat{\mathbf{n}} = \begin{pmatrix} -\sin\varphi \\ \cos\varphi \\ 0 \end{pmatrix}.$$

Thus, we have

$$fI - ig\,\boldsymbol{\sigma}\cdot\hat{\mathbf{n}}$$

$$= \begin{pmatrix} f & 0 \\ 0 & f \end{pmatrix} - i\begin{pmatrix} 0 & -ig\exp(-i\varphi) \\ ig\exp(i\varphi) & 0 \end{pmatrix}$$

$$= i\begin{pmatrix} f & -g\exp(-i\varphi) \\ g\exp(i\varphi) & f \end{pmatrix},$$

(10.59)

so that

$$M(\vartheta,\varphi) = f(\vartheta)\,I - i\,g(\vartheta)\,\boldsymbol{\sigma}\cdot\hat{\mathbf{n}}.$$

(10.60)

To calculate the final polarization, let us first observe that

$$(\boldsymbol{\sigma}\cdot\hat{\mathbf{n}})^\dagger$$

$$= -i\begin{pmatrix} 0 & \exp(-i\varphi) \\ -\exp(i\varphi) & 0 \end{pmatrix}$$

$$= i\begin{pmatrix} 0 & -\exp(-i\varphi) \\ \exp(i\varphi) & 0 \end{pmatrix} = \boldsymbol{\sigma}\cdot\hat{\mathbf{n}},$$

(10.61)

so that

$$(\boldsymbol{\sigma} \cdot \hat{\mathbf{n}})(\boldsymbol{\sigma} \cdot \hat{\mathbf{n}})^\dagger = (\boldsymbol{\sigma} \cdot \hat{\mathbf{n}})(\boldsymbol{\sigma} \cdot \hat{\mathbf{n}})$$

$$= i \begin{pmatrix} 0 & -\exp(-i\varphi) \\ \exp(i\varphi) & 0 \end{pmatrix} i \begin{pmatrix} 0 & -\exp(-i\varphi) \\ \exp(i\varphi) & 0 \end{pmatrix}$$

$$= -\begin{pmatrix} -1 & 0 \\ 0 & -1 \end{pmatrix} = I. \tag{10.62}$$

As a consequence, we have

$$M M^\dagger = (fI - ig\boldsymbol{\sigma} \cdot \hat{\mathbf{n}})[If^* + i(\boldsymbol{\sigma} \cdot \hat{\mathbf{n}})^\dagger g^*]$$
$$= |f|^2 I + |g|^2 (\boldsymbol{\sigma} \cdot \hat{\mathbf{n}})(\boldsymbol{\sigma} \cdot \hat{\mathbf{n}})^\dagger + i(fg^* - f^*g)\boldsymbol{\sigma} \cdot \hat{\mathbf{n}}$$
$$= (|f|^2 + |g|^2)I + i(fg^* - f^*g)\boldsymbol{\sigma} \cdot \hat{\mathbf{n}}$$
$$= (|f|^2 + |g|^2) \begin{pmatrix} 1 & 0 \\ 0 & 1 \end{pmatrix} - (fg^* - f^*g) \begin{pmatrix} 0 & -\exp(-i\varphi) \\ \exp(i\varphi) & 0 \end{pmatrix},$$

or

$$M M^\dagger = \begin{pmatrix} |f|^2 + |g|^2 & (fg^* - f^*g)\exp(-i\varphi) \\ (f^*g - fg^*)\exp(i\varphi) & |f|^2 + |g|^2 \end{pmatrix}. \tag{10.63}$$

Now, it is not difficult to see that

$$M M^\dagger \sigma_x = \begin{pmatrix} (fg^* - f^*g)\exp(-i\varphi) & |f|^2 + |g|^2 \\ |f|^2 + |g|^2 & (f^*g - fg^*)\exp(i\varphi) \end{pmatrix}, \tag{10.64}$$

$$M M^\dagger \sigma_y = i\begin{pmatrix} (fg^* - f^*g)\exp(-i\varphi) & -|f|^2 - |g|^2 \\ |f|^2 + |g|^2 & (fg^* - f^*g)\exp(i\varphi) \end{pmatrix}, \tag{10.65}$$

$$M M^\dagger \sigma_z = \begin{pmatrix} |f|^2 + |g|^2 & (f^*g - fg^*)\exp(-i\varphi) \\ (f^*g - fg^*)\exp(i\varphi) & -|f|^2 - |g|^2 \end{pmatrix}. \tag{10.66}$$

From these equations, it follows that

$$\text{Tr}(M M^\dagger) = 2(|f|^2 + |g|^2), \tag{10.67}$$

$$\text{Tr}(M M^\dagger \sigma_x) = -2i(fg^* - f^*g)\sin\varphi, \tag{10.68}$$

$$\text{Tr}(M M^\dagger \sigma_y) = 2i(fg^* - f^*g)\cos\varphi, \tag{10.69}$$

$$\text{Tr}(M M^\dagger \sigma_z) = 0, \tag{10.70}$$

and, as a consequence,

$$\mathbf{P}_f = \frac{\text{Tr}(M M^\dagger \boldsymbol{\sigma})}{\text{Tr}(M M^\dagger)} = i\frac{fg^* - f^*g}{|f|^2 + |g|^2} \begin{pmatrix} -\sin\varphi \\ \cos\varphi \\ 0 \end{pmatrix} = S(\vartheta)\,\hat{\mathbf{n}}. \tag{10.71}$$

So we have demonstrated that the spin-polarization after scattering is equal, in modulus, to the Sherman function, and it is directed along $\hat{\mathbf{n}}$, i. e., the unit vector normal to the plane of scattering.

10.5 Double elastic scattering

Let us consider now an electron beam initially not polarized experiencing a double scattering. Due to the first elastic collision, the electrons of the beam are deflected with a scattering angle ϑ_1, so that the electrons emerge from this collision with a spin-polarization different from 0. Let us indicate with $\hat{\mathbf{n}}_1$ the unit vector normal to the scattering plane of this first elastic scattering collision. The polarization of the beam after the first elastic scattering collision is given by

$$\mathbf{P}_1 = S(\vartheta_1)\,\hat{\mathbf{n}}_1\,. \tag{10.72}$$

Let us now consider a second elastic scattering collision, with scattering angle ϑ_2. Let $\hat{\mathbf{n}}_2$ be the unit vector normal to the second plane of scattering. The differential elastic scattering cross-section of the second collision is thus given by

$$\begin{aligned}\frac{d\sigma}{d\Omega_2} &= \left(|f(\vartheta_2)|^2 + |g(\vartheta_2)|^2\right)\left(1 + S(\vartheta_2)\mathbf{P}_1\cdot\hat{\mathbf{n}}_2\right)\\ &= \left(|f(\vartheta_2)|^2 + |g(\vartheta_2)|^2\right)\left(1 + S(\vartheta_2)\,S(\vartheta_1)\,\hat{\mathbf{n}}_1\cdot\hat{\mathbf{n}}_2\right).\end{aligned} \tag{10.73}$$

Let us limit ourselves to the case in which the two scattering planes coincide:

$$\hat{\mathbf{n}}_1\cdot\hat{\mathbf{n}}_2 = \pm 1\,, \tag{10.74}$$

so that we have to consider two different differential elastic scattering cross-sections, corresponding to the second elastic collision, depending on the sign of the scalar product $\hat{\mathbf{n}}_1\cdot\hat{\mathbf{n}}_2$. In particular, if $\hat{\mathbf{n}}_1\cdot\hat{\mathbf{n}}_2 = +1$, then the differential elastic scattering cross-section $d\sigma_a/d\Omega_2$ is equal to

$$\frac{d\sigma_a}{d\Omega_2} = \left(|f(\vartheta_2)|^2 + |g(\vartheta_2)|^2\right)\left(1 + S(\vartheta_1)\,S(\vartheta_2)\right), \tag{10.75}$$

while, in the case where $\hat{\mathbf{n}}_1\cdot\hat{\mathbf{n}}_2 = -1$, the differential elastic scattering cross-section $d\sigma_b/d\Omega_2$ is given by

$$\frac{d\sigma_b}{d\Omega_2} = \left(|f(\vartheta_2)|^2 + |g(\vartheta_2)|^2\right)\left(1 - S(\vartheta_1)\,S(\vartheta_2)\right). \tag{10.76}$$

We can thus measure these two differential elastic scattering cross-sections to evaluate the following two quantities:

$$\frac{d\sigma_a}{d\Omega_2} - \frac{d\sigma_b}{d\Omega_2} = 2\left(|f(\vartheta_2)|^2 + |g(\vartheta_2)|^2\right)S(\vartheta_1)\,S(\vartheta_2),\tag{10.77}$$

$$\frac{d\sigma_a}{d\Omega_2} + \frac{d\sigma_b}{d\Omega_2} = 2\left(|f(\vartheta_2)|^2 + |g(\vartheta_2)|^2\right),\tag{10.78}$$

and, hence,

$$S(\vartheta_1)\,S(\vartheta_2) = \frac{d\sigma_a/d\Omega_2 - d\sigma_b/d\Omega_2}{d\sigma_a/d\Omega_2 + d\sigma_b/d\Omega_2}.\tag{10.79}$$

Choosing $\vartheta_1 = \vartheta_2 = \bar{\vartheta}$, it is then possible, with a double scattering experiment, to obtain $S^2(\bar{\vartheta})$. Once $|S(\bar{\vartheta})|$ is known for a given angle $\bar{\vartheta}$, a second experiment is performed varying ϑ_1 and keeping constant $\vartheta_2 = \bar{\vartheta}$. Since $|S(\bar{\vartheta})|$ is known from the first experiment, it is now possible to determine $|S(\vartheta_1)|$ for various angles ϑ_1. By using Eq. (9.70), the sign of $S(\vartheta_1)$ can be found by measuring the differential elastic scattering cross-section of electrons whose polarization direction is known.

10.6 Change of the polarization in the general case

We already know that

$$\mathrm{Tr}(M\,M^\dagger) = 2\left(|f|^2 + |g|^2\right),\tag{10.80}$$

and

$$\mathrm{Tr}(M\,M^\dagger\,\boldsymbol{\sigma}) = 2i(fg^* - f^*g)\,\hat{\mathbf{n}}.\tag{10.81}$$

Furthermore, we can see that[5]

$$\mathrm{Tr}(M\,(\boldsymbol{\sigma}\cdot\mathbf{P})\,M^\dagger) = 2i(fg^* - f^*g)\,\mathbf{P}\cdot\hat{\mathbf{n}}\tag{10.82}$$

and

$$\mathrm{Tr}(M\,(\boldsymbol{\sigma}\cdot\mathbf{P})\,M^\dagger\,\boldsymbol{\sigma}) = 2\left(|f|^2 - |g|^2\right)\mathbf{P} - 2\left(fg^* + f^*g\right)\mathbf{P}\times\hat{\mathbf{n}} + 4|g|^2\,(\mathbf{P}\cdot\hat{\mathbf{n}})\,\hat{\mathbf{n}}.\tag{10.83}$$

Let us introduce now two further polarization parameters, i. e.,

$$T(\vartheta) = \frac{|f(\vartheta)|^2 - |g(\vartheta)|^2}{|f(\vartheta)|^2 + |g(\vartheta)|^2},\tag{10.84}$$

---

[5] The demonstration of these two equations, however lengthy and laborious, is not difficult. It is left as a useful exercise for the reader to solve.

and

$$U(\theta) = \frac{f(\vartheta)g^*(\vartheta) + f^*(\vartheta)g(\vartheta)}{|f(\vartheta)|^2 + |g(\vartheta)|^2}. \tag{10.85}$$

By using Eq. (10.56), and taking into account the relationship:

$$\frac{2|g|^2}{|f|^2 + |g|^2} = 1 - T(\vartheta), \tag{10.86}$$

we obtain, in the general case $0 \le |\mathbf{P}| \le 1$,

$$\mathbf{P}_f = \frac{[\mathbf{P} \cdot \hat{\mathbf{n}} + S(\vartheta)]\,\hat{\mathbf{n}} + T(\vartheta)[\mathbf{P} - (\mathbf{P} \cdot \hat{\mathbf{n}})\,\hat{\mathbf{n}}] + U(\vartheta)\,(\hat{\mathbf{n}} \times \mathbf{P})}{1 + (\mathbf{P} \cdot \hat{\mathbf{n}})\,S(\vartheta)}. \tag{10.87}$$

Note that

$$S^2(\vartheta) + T^2(\vartheta) + U^2(\vartheta) = 1. \tag{10.88}$$

10.7 Sherman function for molecules

Let us firstly consider oriented molecules. In this case the spin-asymmetry function (Sherman function) is given by

$$S(\vartheta) = \frac{i}{d\sigma/d\Omega} \sum_m \sum_n [\exp(i\mathbf{q} \cdot \mathbf{r}_{mn})f_m(\vartheta)g_n^*(\vartheta) - \exp(-i\mathbf{q} \cdot \mathbf{r}_{mn})f_m^*(\vartheta)g_n(\vartheta)], \tag{10.89}$$

where

$$\frac{d\sigma}{d\Omega} = \sum_m \sum_n \exp(i\mathbf{q} \cdot \mathbf{r}_{mn})[f_m(\vartheta)f_n^*(\vartheta) + g_m(\vartheta)g_n^*(\vartheta)].$$

Please keep in mind that $\mathbf{q}$ is the momentum transfer, $\mathbf{r}_m$ is the position vector of the mth atom in the molecule, and

$$\mathbf{r}_{mn} = \mathbf{r}_m - \mathbf{r}_n.$$

In case the molecules are randomly oriented, these equations simplify and become

$$S(\vartheta) = \frac{i}{d\sigma/d\Omega} \sum_m \sum_n \frac{\sin qr_{mn}}{qr_{mn}}[f_m(\vartheta)g_n^*(\vartheta) - f_m^*(\vartheta)g_n(\vartheta)], \tag{10.90}$$

where

$$\frac{d\sigma}{d\Omega} = \sum_m \sum_n \frac{\sin qr_{mn}}{qr_{mn}}[f_m(\vartheta)f_n^*(\vartheta) + g_m(\vartheta)g_n^*(\vartheta)],$$

and

$$r_{mn} = |\mathbf{r}_{mn}| = |\mathbf{r}_m - \mathbf{r}_n|. \qquad (10.91)$$

Part IV: **Applications**

11 Exercises

11.1 Exercise 1

Using their analytical expressions, calculate and represent the first five Legendre polynomials $P_l(\cos \vartheta)$.

Solution

We know that

$$P_0(u) = 1, \tag{11.1}$$
$$P_1(u) = u, \tag{11.2}$$
$$P_2(u) = \frac{1}{2}(3u^2 - 1), \tag{11.3}$$
$$P_3(u) = \frac{1}{2}(5u^3 - 3u), \tag{11.4}$$
$$P_4(u) = \frac{1}{8}(35u^4 - 30u^2 + 3). \tag{11.5}$$

In Fig. 11.1 we represent the first five Legendre polynomials obtained using these equations with $u = \cos \vartheta$.

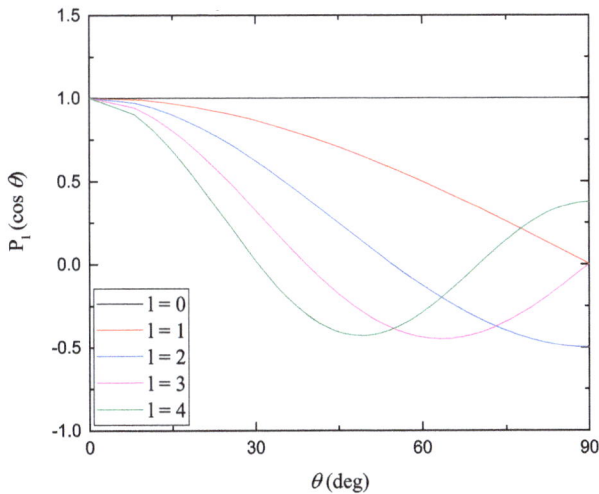

Figure 11.1: The first five Legendre polynomials calculated using their analytical expressions.

https://doi.org/10.1515/9783110675375-011

11.2 Exercise 2

Using recursion relations, numerically calculate and represent the first six Legendre polynomials $P_l(\cos\vartheta)$.

Solution

The Legendre polynomials $P_l(u)$ satisfy the following recursion relation:

$$(l+1)\,P_{l+1}(u) + l\,P_{l-1}(u) = (2l+1)\,u\,P_l(u)\,. \tag{11.6}$$

Since we know that

$$P_0(u) = 1\,, \tag{11.7}$$
$$P_1(u) = u\,, \tag{11.8}$$

we can apply a forward recursion procedure. In particular, we will then have

$$2\,P_2(u) = -P_0(u) + 3\,u\,P_1(u)\,, \tag{11.9}$$
$$3\,P_3(u) = -2\,P_1(u) + 5\,u\,P_2(u)\,, \tag{11.10}$$
$$4\,P_4(u) = -3\,P_2(u) + 7\,u\,P_3(u)\,, \tag{11.11}$$
$$5\,P_5(u) = -4\,P_3(u) + 9\,u\,P_4(u)\,, \tag{11.12}$$

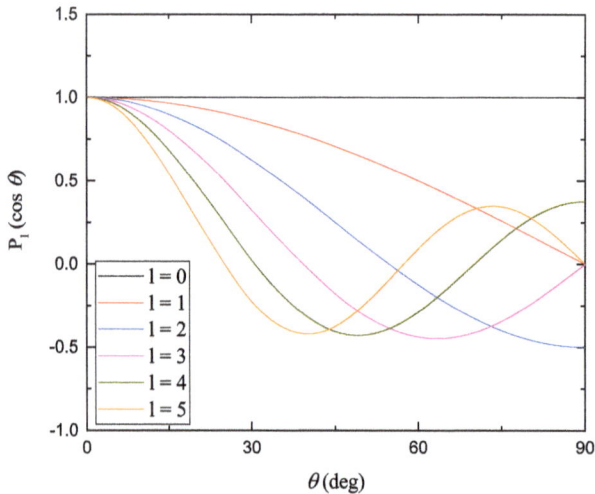

Figure 11.2: The first six Legendre polynomials $P_l(\cos\vartheta)$ numerically calculated using the recursion relation.

and so on. In Fig. 11.2 we represent the first six Legendre polynomials obtained using these recursion relations with $u = \cos\vartheta$.

11.3 Exercise 3

Numerically calculate and represent the first six associated Legendre functions $P_l^1(\cos\vartheta)$.

Solution

The first derivative of the Legendre polynomials can be evaluated using the following equation:

$$(1 - u^2)\frac{d}{du}P_l(u) = l\,P_{l-1}(u) - l\,u\,P_l(u). \tag{11.13}$$

Knowing the Legendre polynomials, it is possible, as a consequence, to calculate the associated Legendre functions that, as a function of $u = \cos\vartheta$, are given by

$$P_l^1(\cos\vartheta) = \sin\vartheta\frac{d}{d(\cos\vartheta)}P_l(\cos\vartheta). \tag{11.14}$$

In Fig. 11.3 we represent the first six associated Legendre functions obtained using Eqs. (11.13) and (11.14).

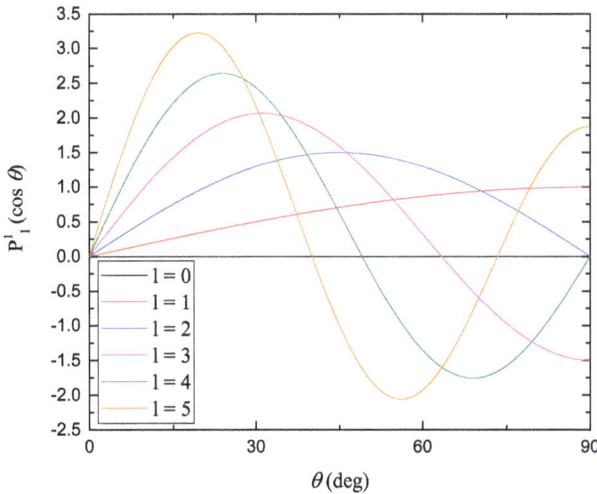

Figure 11.3: The first six associated Legendre functions $P_l^1(\cos\vartheta)$ numerically calculated using Eq. (11.13).

11.4 Exercise 4

Analytically calculate, using their explicit expressions, the associated Legendre functions $P_3^1(0.5)$ and $P_4^1(0.5)$.

Solution

The associated Legendre functions $P_l^1(u)$ are given by

$$P_l^1(u) = \sqrt{1 - u^2} \frac{d}{du} P_l(u). \tag{11.15}$$

Since

$$\frac{dP_3(u)}{du} = \frac{1}{2}(15u^2 - 3), \tag{11.16}$$

we have

$$\frac{dP_3(0.5)}{du} = \frac{3}{8} \tag{11.17}$$

and

$$P_3^1(0.5) = \sqrt{1 - \frac{1}{4}} \frac{3}{8} = \frac{3}{16} \sqrt{3}. \tag{11.18}$$

Let us now calculate $P_4^1(0.5)$. From

$$\frac{dP_4(u)}{du} = \frac{1}{8}(140u^3 - 60u), \tag{11.19}$$

it follows that

$$\frac{dP_4(0.5)}{du} = -\frac{25}{16} \tag{11.20}$$

and, hence,

$$P_4^1(0.5) = -\sqrt{1 - \frac{1}{4}} \frac{25}{16} = -\frac{25}{32} \sqrt{3}. \tag{11.21}$$

11.5 Exercise 5

Demonstrate that

$$4 P_4(u) = -3 P_2(u) + 7 u P_3(u), \tag{11.22}$$

where

$$P_4(u) = \frac{1}{8}(35u^4 - 30u^2 + 3).$$ (11.23)

Solution

We know that

$$P_2(u) = \frac{1}{2}(3u^2 - 1),$$ (11.24)

$$P_3(u) = \frac{1}{2}(5u^3 - 3u).$$ (11.25)

As a consequence,

$$-3P_2(u) + 7uP_3(u)$$

$$= -3\left[\frac{1}{2}(3u^2 - 1)\right] + 7u\left[\frac{1}{2}(5u^3 - 3u)\right]$$

$$= -\frac{9u^2}{2} + \frac{3}{2} + \frac{35u^4}{2} - \frac{21u^2}{2}$$

$$= 4\frac{1}{8}(35u^4 - 30u^2 + 3) = 4P_4(u).$$

Note that this is an application of the recursion relation Eq. (11.6).

11.6 Exercise 6

Calculate $P_5(u)$.

Solution

Using the recursion relation Eq. (11.6), we obtain:

$$P_5(u)$$

$$= \frac{1}{5}[-4P_3(u) + 9uP_4(u)]$$

$$= \frac{1}{5}\left[-2(5u^3 - 3u) + \frac{9}{8}u(35u^4 - 30u^2 + 3)\right]$$

$$= \frac{1}{8}(63u^5 - 70u^3 + 15u).$$ (11.26)

11.7 Exercise 7

Demonstrate that

$$n_2(x) = \left(-\frac{3}{x^3} + \frac{1}{x}\right)\cos x - \frac{3}{x^2}\sin x. \tag{11.27}$$

Solution

Let us remind our readers that the following recursion relation holds:

$$x\, n_{l-1}(x) - (2l + 1)\, n_l(x) + x\, n_{l+1}(x) = 0, \tag{11.28}$$

and that

$$n_0(x) = -\frac{\cos x}{x}, \tag{11.29}$$

$$n_1(x) = -\frac{\cos x}{x^2} - \frac{\sin x}{x}. \tag{11.30}$$

Using Eq. (11.28) with $l = 1$, we obtain

$$n_2(x) = \frac{3}{x}\, n_1(x) - n_0(x), \tag{11.31}$$

and, as a consequence,

$$n_2(x) = \frac{3}{x}\left(-\frac{\cos x}{x^2} - \frac{\sin x}{x}\right) + \frac{\cos x}{x} = \left(-\frac{3}{x^3} + \frac{1}{x}\right)\cos x - \frac{3}{x^2}\sin x. \tag{11.32}$$

11.8 Exercise 8

Demonstrate that

$$j_0(x) = \frac{\sin x}{x}. \tag{11.33}$$

Solution

The recursion relation:

$$x\, j_{l-1}(x) - (2l + 1)\, j_l(x) + x\, j_{l+1}(x) = 0, \tag{11.34}$$

allows us to see, on the one hand, that, once $l = 1$ is chosen,

$$j_0(x) = \frac{3}{x} j_1(x) - j_2(x).$$

(11.35)

Since, on the other hand,

$$j_1(x) = \frac{\sin x}{x^2} - \frac{\cos x}{x},$$

(11.36)

$$j_2(x) = \left(\frac{3}{x^3} - \frac{1}{x}\right) \sin x - \frac{3}{x^2} \cos x,$$

(11.37)

we have

$$j_0(x) = \frac{3}{x}\left(\frac{\sin x}{x^2} - \frac{\cos x}{x}\right) - \left(\frac{3}{x^3} - \frac{1}{x}\right)\sin x + \frac{3}{x^2}\cos x = \frac{\sin x}{x}.$$

(11.38)

11.9 Exercise 9

Express analytically $n_3(x)$ and $j_3(x)$.

Solution

Using Eq. (11.28) with $l = 2$, we obtain

$$n_3(x) = \frac{5}{x} n_2(x) - n_1(x),$$

(11.39)

so that

$$n_3(x)$$
$$= \frac{5}{x}\left[\left(-\frac{3}{x^3} + \frac{1}{x}\right)\cos x - \frac{3}{x^2}\sin x\right] + \frac{\cos x}{x^2} + \frac{\sin x}{x}$$
$$= \left(-\frac{15}{x^4} + \frac{6}{x^2}\right)\cos x + \left(-\frac{15}{x^3} + \frac{1}{x}\right)\sin x.$$

(11.40)

Similarly, using Eq. (11.34) with $l = 2$, we obtain

$$j_3(x) = \frac{5}{x} j_2(x) - j_1(x),$$

(11.41)

so that

$$j_3(x)$$

$$= \frac{5}{x}\left[\left(\frac{3}{x^3} - \frac{1}{x}\right)\sin x - \frac{3}{x^2}\cos x\right] - \frac{\sin x}{x^2} + \frac{\cos x}{x}$$

$$= \left(\frac{15}{x^4} - \frac{6}{x^2}\right)\sin x + \left(-\frac{15}{x^3} + \frac{1}{x}\right)\cos x. \tag{11.42}$$

11.10 Exercise 10

Write a program to calculate $n_l(x)$ for any given value of l.

Solution

Let us use a forward recursion method. Since we know $n_0(x)$ and $n_1(x)$, we can calculate $n_l(x)$ just by recursively using Eq. (11.28). In Fig. 11.4 we represent the numerically calculated first four irregular spherical Bessel functions recursively calculated in such a way.

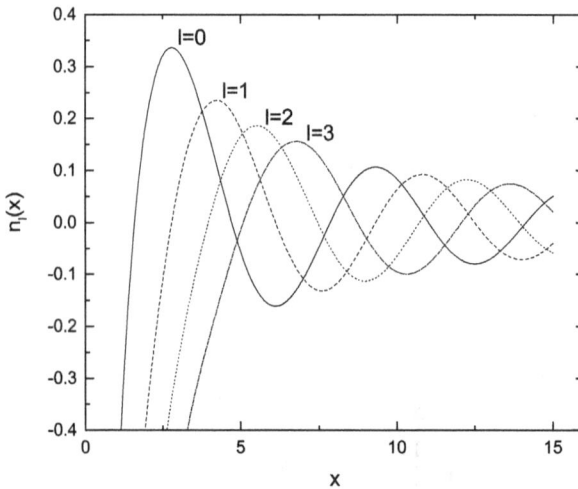

Figure 11.4: The first four irregular Bessel functions numerically calculated using a forward recursion algorithm. The program utilizes Eqs. (11.28), (11.29), and (11.30).

11.11 Exercise 11

Write a program to calculate $j_l(x)$ for any given value of l.

Solution

Let us use a backward recursion method imposing, to start the recursion, that $j_{l_{max}} = 0$ and $j_{l_{max}-1} = \epsilon$, where l_{max} is a very large value of l and ϵ is a very small number. We then calculate $j_l(x)$ recursively using Eq. (11.34) and utilizing a backward procedure. Note that the results obtained with this backward procedure have to be normalized so that $j_0(x) = \sin x/x$. In Fig. 11.5 we represent the numerically calculated first four regular spherical Bessel functions recursively calculated in such a way.

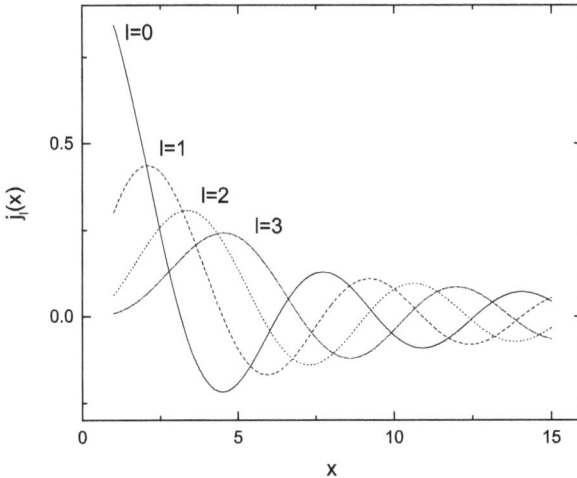

Figure 11.5: The first four regular Bessel functions numerically calculated using a backward recursion algorithm. The program utilizes Eq. (11.34). Eq. (11.33) is used to normalize the results.

11.12 Exercise 12

Discuss the differences among forward and backward recursion procedures for calculating the spherical Bessel functions.

Solution

The question we wish to discuss concerns the correct use of forward and backward recursion procedures. Let us consider the numerical differentiation method. We know that, since

$$f_1 \equiv f(x + h) = f_0 + hf' + \frac{h^2}{2}f'' + \frac{h^3}{6}f''' + \cdots \tag{11.43}$$

and

$$f_{-1} \equiv f(x - h) = f_0 - hf' + \frac{h^2}{2}f'' - \frac{h^3}{6}f''' + \cdots, \tag{11.44}$$

we have

$$f_1 - 2f_0 + f_{-1} = h^2 f'' + \mathcal{O}(h^4), \tag{11.45}$$

so that

$$f'' \approx \frac{f_1 - 2f_0 + f_1}{h^2}. \tag{11.46}$$

Please note now that the recursion relation, satisfied by both the irregular and the regular spherical Bessel functions, can be rewritten as follows:

$$f_{l+1} - 2f_l + f_{l-1} = \left[\frac{2l+1}{x} - 2\right]f_l, \tag{11.47}$$

so that, in the limit of continuous l [$f_l = f(l)$], we can write

$$\frac{d^2 f(l)}{dl^2} = \frac{2l - 2x + 1}{x} f(l) = -k^2(l)f(l). \tag{11.48}$$

This equation, in the case $k^2(l) = (2x - 2l - 1)/x > 0$, is satisfied by two oscillatory linearly independent functions while, if $k^2(l) < 0$, it is satisfied by two exponential linearly independent functions, one growing and the other one decreasing. In general, when

$$x < \frac{2l+1}{2}, \tag{11.49}$$

and, as a consequence,

$$k^2(l) < 0, \tag{11.50}$$

the regular spherical Bessel functions $j_l(x)$ rapidly decrease with increasing l, so that precision is lost using forward recursion in calculating $j_l(x)$. For this reason, regular spherical Bessel functions $j_l(x)$ have to be numerically calculated using backward recursion. As a general rule, we should avoid recursion in the same direction in which the values of the function we are calculating become smaller and smaller. So, while forward recursion can be used to calculate the irregular spherical Bessel functions

$n_l(x)$, this method has to be avoided when calculating the regular spherical Bessel functions $j_l(x)$. Please note that backward recursion in calculating $j_l(x)$ provides more accurate results if compared with those obtained using both forward recursion formulas and explicit evaluation.

11.13 Exercise 13

Write a program to calculate the differential elastic scattering cross-section of 1,000 eV electrons in Au using the Schrödinger equation (nonrelativistic partial wave expansion method). Use the Numerov algorithm for solving the radial equation.

Solution

We have to solve the radial equation:

$$\frac{d^2 F_l(r)}{dr^2} + q^2(r) F_l(r) = 0, \tag{11.51}$$

where $q^2(r)$ depends on l, on the electron energy E, and on the atomic potential $V(r)$ according to Eq. (5.1). We knows that the Numerov algorithm allows us to find $F_l(r)$ by numerically solving the following equation:

$$\left(1 + \frac{h^2}{12} q_{n+1}^2\right) (F_l)_{n+1}$$
$$- 2\left(1 - \frac{5 h^2}{12} q_n^2\right) (F_l)_n$$
$$+ \left(1 + \frac{h^2}{12} q_{n-1}^2\right) (F_l)_{n-1} \approx 0. \tag{11.52}$$

To solve it, we need, of course, to know the potential. So, the first step is to write a program that calculates the atomic potential energy $V(r)$. Let us choose, for this exercise, the Lenz and Jensen approximation to the Thomas–Fermi model of the atom:

$$V(r) = -\frac{Ze^2}{r} \xi(x), \tag{11.53}$$

where

$$\xi(x) = \exp(-x)(1 + x + ax^2 + bx^3 + cx^3), \tag{11.54}$$
$$x = 4.5397 \, Z^{1/6} \sqrt{r}, \tag{11.55}$$

and $a = 0.3344$, $b = 0.0485$, and $c = 0.002647$. Please note that the unit of length is Å, using the numerical parameters here provided [15]. Once the potential is known, the second step requires that the radial function $F_l(r)$ be calculated for two values of r, r_1, and r_2, both greater than $r_{max} = 3$ Å (the radius beyond which the potential becomes negligible). When choosing r_1 and r_2, make sure that they are not too close to each other (to obtain numerical precision). On the other hand, the greater the distance between r_1 and r_2, the longer the CPU time necessary for the Numerov algorithm to process the entire computation.

The third step consists of computing, for each l between 0 and l_{max} (according to Koonin and Meredith [15], a slightly low estimate of it is $l_{max} \approx k r_{max}$) the quantity:

$$G(r_1, r_2) = \frac{r_1 F_l(r_2)}{r_2 F_l(r_1)}. \tag{11.56}$$

Thus, using the programs already written for calculating the spherical regular and irregular Bessel functions, the fourth step consists of writing a code for calculating the phase shifts given by

$$\eta_l = \arctan \frac{G(r_1, r_2) j_l(kr_1) - j_l(kr_2)}{G(r_1, r_2) n_l(kr_1) - n_l(kr_2)}. \tag{11.57}$$

Once the phase shifts have been calculated, the last step requires the use of the programs already written for computing the Legendre polynomials to calculate the scattering amplitude according to Eq. (3.109) and the differential elastic scattering

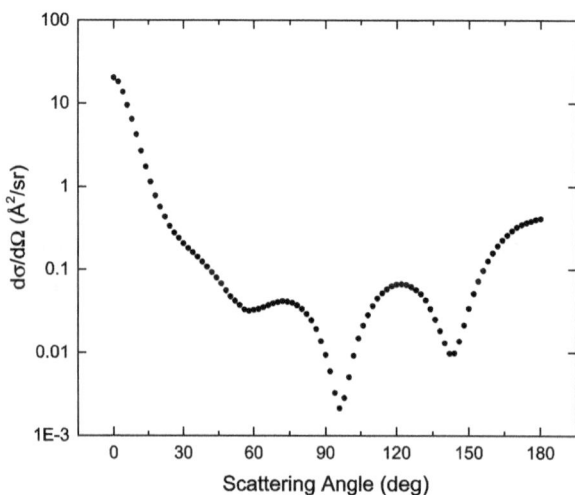

Figure 11.6: Differential elastic scattering cross-section of 1,000 eV electrons calculated for Au using the Schrödinger equation (nonrelativistic partial wave expansion method). Atomic potential: Lenz and Jensen approximation to the Thomas–Fermi model of the atom.

cross-section according to Eq. (3.111). The differential elastic scattering cross-section of 1,000 eV electrons, calculated for Au using the procedure previously described, is shown in Fig. 11.6.

11.14 Exercise 14

Write a program to calculate the screening functions of Al, Ag, and Au using the Salvat et al. best fit of the Dirac–Hartree–Fock–Slater model of the atom [24]. Use atomic units: $\hbar = e = m = 1$.

Solution

The atomic potential energy can be calculated as the product of the Coulomb potential energy multiplied by a screening function. Salvat et al. [24] provided the following screening function representing the best fit of their Dirac–Hartree–Fock–Slater calculations:

$$\xi(r) = \sum_{i=1}^{3} \gamma_i \exp[-\lambda_i r]. \qquad (11.58)$$

The values of the parameters γ_i and λ_i (with $i = 1, 2, 3$) of Al, Ag, and Au, according to Salvat et al., are given in Table 11.1 [24]. Note that the numerical values of λ_i given in Table 11.1 are expressed in atomic units, so that $\hbar = e = m = 1$ and the unit of length, as a consequence, is the Bohr radius $a_0 = \hbar^2/me^2$. Using the parameters in Table 11.1, the screening functions of Al, Ag, and Au are represented in Fig. 11.7 as a function of the distance r expressed in units a_0.

Table 11.1: γ_i and λ_i ($i = 1, 2, 3$) parameters of Al, Ag, and Au according to the Salvat et al. best fit of the Dirac–Hartree–Fock–Slater potential energy [24].

Z	γ_1	γ_2	γ_3	λ_1	λ_2	λ_3
13	0.6002	0.3998	0.0000	5.1405	1.0153	0.0000
47	0.2562	0.6505	0.0933	15.588	2.7412	1.1408
79	0.2289	0.6114	0.1597	22.864	3.6914	1.4886

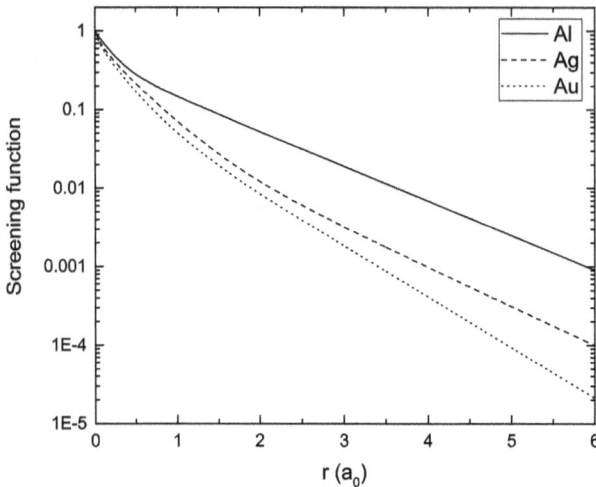

Figure 11.7: Screening functions of Al, Ag, and Au as a function of the distance r expressed in units a_0. Salvat et al. best fit of the of the Dirac–Hartree–Fock–Slater atomic model [24].

11.15 Exercise 15

Write a program to calculate the screening functions of Al, Ag, and Au using the Lenz and Jensen best fit of the Thomas–Fermi atomic model. Compare with the same calculations performed using the Salvat et al. best fit of the Dirac–Hartree–Fock–Slater model of the atom (previous exercise). Use atomic units: $\hbar = e = m = 1$.

Solution

The comparison between the screening functions of Al, Ag, and Au obtained by using the two models is presented in Fig. 11.8 as a function of the distance r expressed in units a_0.

11.16 Exercise 16

Calculate the screening functions of Al and Ag using the Cox and Bonham parameters [7]. Compare with the screening functions of Al and Ag obtained utilizing the Salvat et al. parameters [24]. Use atomic units: $\hbar = e = m = 1$.

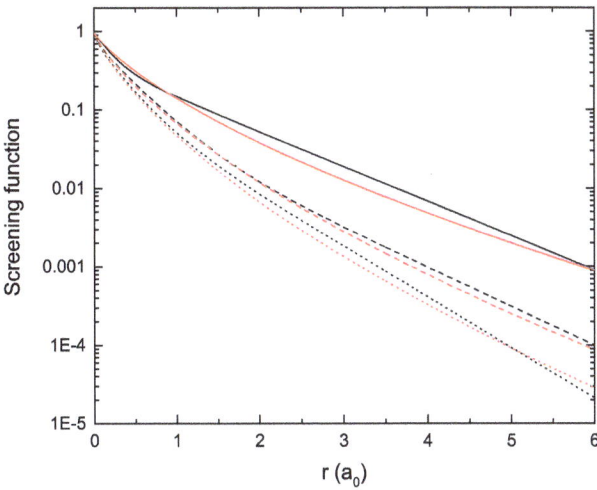

Figure 11.8: Screening functions of Al (solid lines), Ag (dashed lines), and Au (dotted lines) as a function of the distance r expressed in units a_0. Comparison between the Lenz and Jensen best fit of the Thomas–Fermi (red lines) [15] and the Salvat et al. best fit of the Dirac–Hartree–Fock–Slater (black lines) [24] atomic models.

Solution

The Cox and Bonham [7] screening function is given by

$$\xi(r) = \sum_{i=1}^{10} \gamma_i \exp[-\lambda_i r], \tag{11.59}$$

where the parameters γ_i and λ_i (atomic units) for Al and Ag are given in Tables 11.2 and 11.3 ($i = 1, \ldots, 10$). In Fig. 11.9 we show the comparison between the screening

Table 11.2: γ_i ($i = 1, \ldots, 10$) parameters of Al and Ag according to Cox and Bonham [7].

Z	γ_1	γ_2	γ_3	γ_4	γ_5	γ_6	γ_7	γ_8	γ_9	γ_{10}
13	0.6481	−0.0730	2.5058	4.4794	−1.7928	−4.7699	0.0000	0.0000	0.0000	0.0000
47	6.2210	−0.0303	18.1510	18.1633	−3.8052	−21.4036	−19.6939	1.4843	1.7435	0.1594

Table 11.3: λ_i ($i = 1, \ldots, 10$) parameters of Al and Ag according to Cox and Bonham [7].

Z	λ_1	λ_2	λ_3	λ_4	λ_5	λ_6	λ_7	λ_8	λ_9	λ_{10}
13	1.1459	34.0620	3.5958	12.6503	2.6209	11.7028	0.0000	0.0000	0.0000	0.0000
47	11.2381	121.1716	19.5404	42.3580	5.7667	17.1382	41.2617	6.9957	3.6234	1.4254

functions of Al and Ag calculated using Cox and Bonham [7] and Salvat et al. [24] fitting parameters.

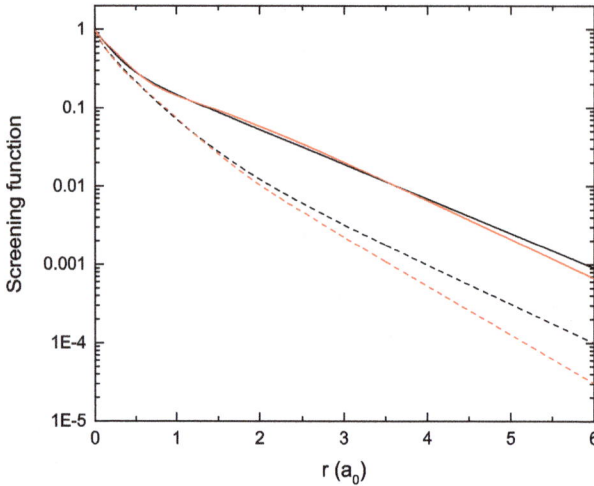

Figure 11.9: Screening functions of Al (solid lines) and Ag (dashed lines) as a function of the distance r expressed in units a_0. Comparison between the calculations obtained using Cox and Bonham (red lines) [7] and Salvat et al. (black lines) [24] best fit parameters.

11.17 Exercise 17

Calculate the radial atomic density of Ar, Kr, and Xe using the Cox and Bonham screening function [7].

Solution

The Cox and Bonham parameters y_i and λ_i (atomic units) for Ar, Kr, and Xe are given in Tables 11.4 and 11.5 ($i = 1, \ldots, 10$). From Poisson's equation, it follows that the radial atomic density is given by

$$\rho(r) = \frac{Z}{4\pi r} \sum_{i=1}^{10} y_i \lambda_i^2 \exp[-\lambda_i r].$$
(11.60)

Using Eq. (11.60) and Tables 11.4 and 11.5, we have represented the radial atomic densities of Ar, Kr, and Xe in Figs. 11.10, 11.11, and 11.12, respectively.

Table 11.4: γ_i ($i = 1,\ldots,10$) parameters of Ar, Kr, and Xe according to Cox and Bonham [7].

Z	γ_1	γ_2	γ_3	γ_4	γ_5	γ_6	γ_7	γ_8	γ_9	γ_{10}
18	1.4268	−0.0602	4.6440	7.4701	−4.4056	−8.0771	0.0000	0.0000	0.0000	0.0000
36	4.6222	−0.0484	12.8497	14.6739	−1.6424	−15.3978	−15.5905	1.5224	0.0000	0.0000
54	7.6397	−0.0299	21.5995	20.8424	−5.1302	−25.0317	−22.8271	1.5368	2.1983	0.2210

Table 11.5: λ_i ($i = 1,\ldots,10$) parameters of Ar, Kr, and Xe according to Cox and Bonham [7].

Z	λ_1	λ_2	λ_3	λ_4	λ_5	λ_6	λ_7	λ_8	λ_9	λ_{10}
18	2.1236	46.3176	5.7689	16.7556	4.5419	15.8867	0.0000	0.0000	0.0000	0.0000
36	7.2780	85.3334	13.1155	33.2254	2.8197	11.3682	32.3498	2.0998	0.0000	0.0000
54	13.6666	137.4095	24.0007	48.7608	7.8087	21.1625	47.4972	9.1558	4.9962	0.8904

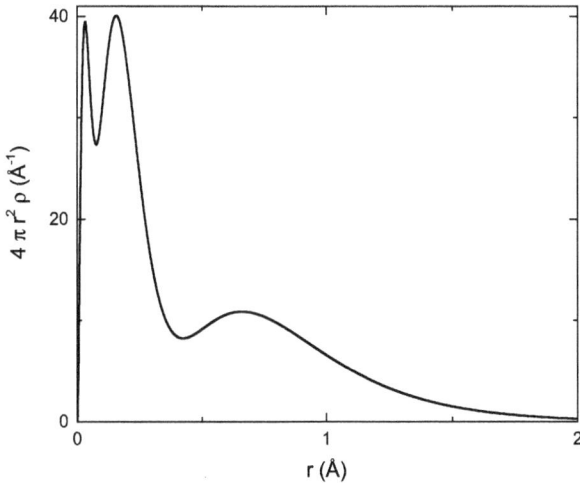

Figure 11.10: Radial atomic density of Ar calculated using the Cox and Bonham screening function [7].

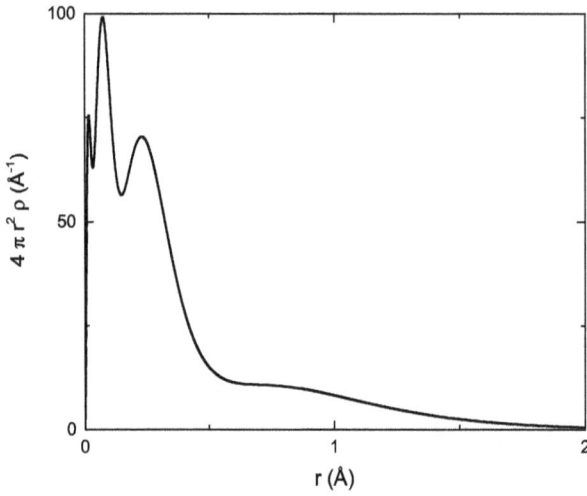

Figure 11.11: Radial atomic density of Kr calculated using the Cox and Bonham screening function [7].

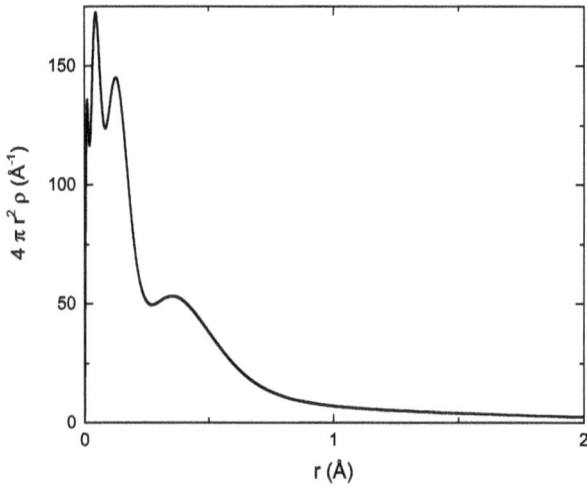

Figure 11.12: Radial atomic density of Xe calculated using the Cox and Bonham screening function [7].

11.18 Exercise 18

Calculate the radial atomic density of Ar, Kr, and Xe using the Salvat et al. screening function [24].

Solution

The Salvat et al. parameters y_i and λ_i (atomic units) for Ar, Kr, and Xe are given in Table 11.6 ($i = 1, 2, 3$). From Poisson's equation, it follows that the radial atomic density is given by

$$\rho(r) = \frac{Z}{4\pi r} \sum_{i=1}^{3} y_i \lambda_i^2 \exp[-\lambda_i r] .$$

(11.61)

Using Eq. (11.61) and Table 11.6, we have represented the radial atomic densities of Ar, Kr, and Xe in Figs. 11.13, 11.14, and 11.15, respectively. Not surprisingly, the analytical screening function of Salvat et al., considering just two or three Yukawa potentials, only partially reproduces the oscillations of the radial atomic density due to the different shell contributions [24].

Table 11.6: y_i and λ_i ($i = 1, 2, 3$) parameters of Ar, Kr, and Xe according to Salvat et al. [24].

Z	Y_1	Y_2	Y_3	λ_1	λ_2	λ_3
18	2.1912	−2.2852	1.0940	5.5470	4.5687	2.0446
36	0.4190	0.5810	0.0000	9.9142	1.8835	0.0000
54	0.4451	0.5549	0.0000	11.805	1.7967	0.0000

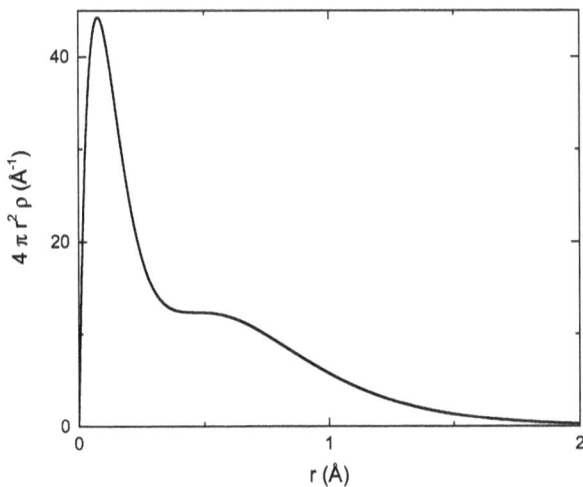

Figure 11.13: Radial atomic density of Ar calculated using the Salvat et al. screening function [24].

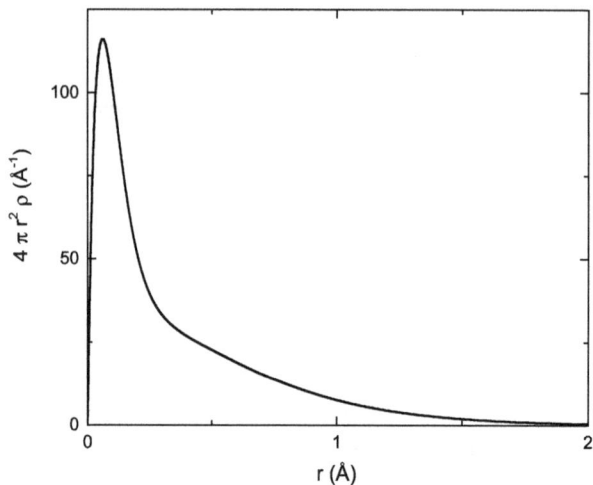

Figure 11.14: Radial atomic density of Kr calculated using the Salvat et al. screening function [24].

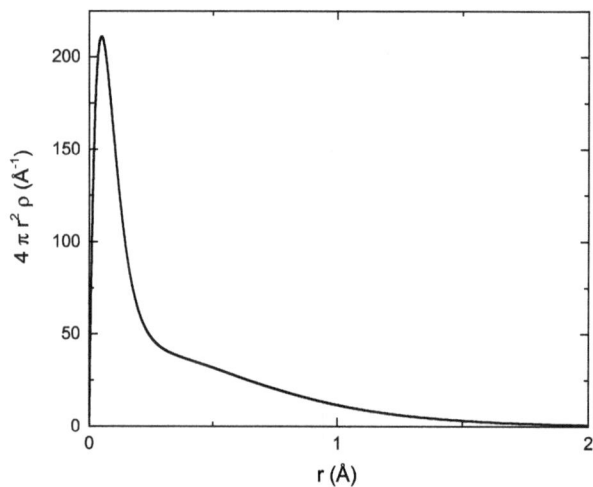

Figure 11.15: Radial atomic density of Xe calculated using the Salvat et al. screening function [24].

11.19 Exercise 19

Calculate the phase shifts for 1,000 eV electrons in Al using the Mott theory. Compute the atomic potential utilizing the Cox and Bonham screening function. Demonstrate that, apart from the initial behavior, the phase shifts go quickly to zero as l increases.

Solution

To calculate the phase shifts given by:

$$\tan \eta_l^{\pm} = \frac{Kj_{l+1}(Kr) - j_l(Kr)[(E+m)\tan \tilde{\phi}_l^{\pm} + (1+l+k)/r]}{Kn_{l+1}(Kr) - n_l(Kr)[(E+m)\tan \tilde{\phi}_l^{\pm} + (1+l+k)/r]},$$ (11.62)

it is necessary to calculate $\tilde{\phi}_l^{\pm}$, where

$$\tilde{\phi}_l^{\pm} = \lim_{r \to \infty} \phi_l^{\pm}(r).$$ (11.63)

The functions $\phi_l^{\pm}(r)$ are the solutions to the differential equation:

$$\frac{d\phi_l^{\pm}(r)}{dr} = \frac{k}{r}\sin 2\phi_l^{\pm}(r) - m\cos 2\phi_l^{\pm}(r) + E - V(r).$$ (11.64)

that can be numerically solved using the fourth-order Runge–Kutta method. The initial value of $\phi_l^{\pm}$, necessary to start the numerical solution of the differential equation, are given by Eqs. (9.108)–(9.111). Using this procedure, the first ≈ 60 phase shifts for 1,000 eV electrons in Al are presented in Figs. 11.16 (η_+) and 11.17 (η_-).

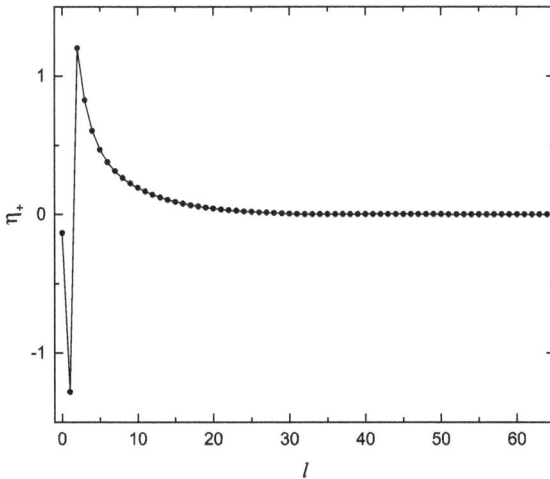

Figure 11.16: η_+ as a function of l for 1,000 eV electrons in Al according to the Mott theory (symbols). Cox and Bonham screening function [7]. Solid line is a guide for the eyes.

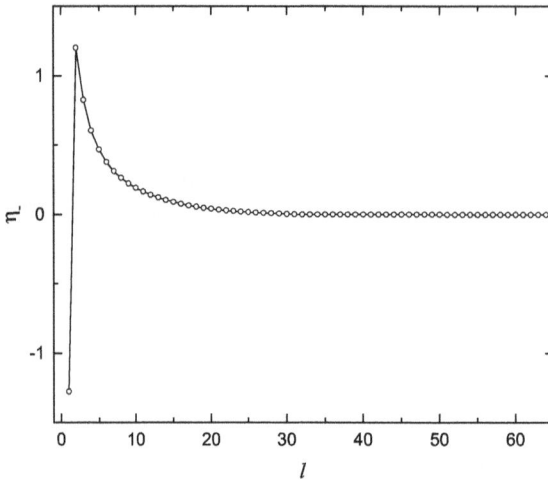

Figure 11.17: η_- as a function of l for 1,000 eV electrons in Al according to the Mott theory (symbols). Cox and Bonham screening function [7]. Solid line is a guide for the eyes.

11.20 Exercise 20

Calculate the differential elastic scattering cross-section for 1,000 eV electrons in Au using the Mott theory. Compute the atomic potential using the Salvat et al. screening function [24].

Solution

According to the Mott theory, once phase shifts η_+ and η_- have been calculated, the scattering amplitudes can be obtained by

$$f(\vartheta) = \frac{1}{2iK} \sum_{l=0}^{\infty} \{(l+1)[\exp(2i\eta_l^+) - 1] + l[\exp(2i\eta_l^-) - 1]\} P_l(\cos\vartheta), \qquad (11.65)$$

$$g(\vartheta) = \frac{1}{2iK} \sum_{l=0}^{\infty} \{\exp(2i\eta_l^-) - \exp(2i\eta_l^+)\} P_l^1(\cos\vartheta), \qquad (11.66)$$

and the differential elastic scattering cross-section, for a completely unpolarized electron beam, can be obtained by

$$\frac{d\sigma}{d\Omega} = |f|^2 + |g|^2. \qquad (11.67)$$

The differential elastic scattering cross-section of 1,000 eV electrons, calculated for Au using the previously described procedure, is shown in Fig. 11.18.

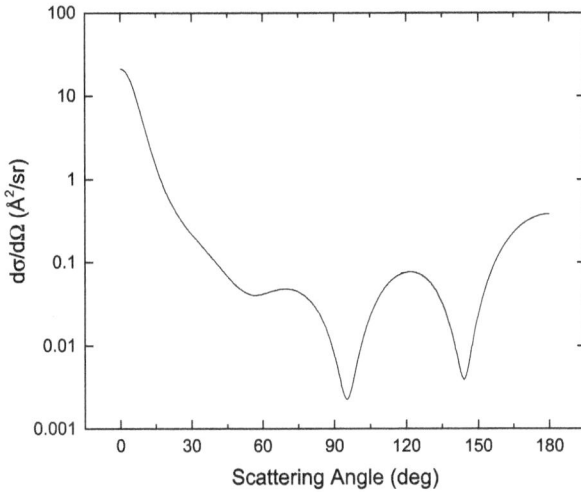

Figure 11.18: The differential elastic scattering cross-section of 1,000 eV electrons calculated for Au using the Dirac equation (relativistic partial wave expansion method). The Salvat et al. screening function [24] was used to calculate the electrostatic potential.

11.21 Exercise 21

Calculate the differential elastic scattering cross-section for 1,000 eV electrons in H_2O using the Mott theory. Compute the atomic potential using the Cox and Bonham screening function [7].

Solution

Once phase shifts η_+ and η_- have been calculated for H and O using the Cox and Bonham atomic potential [see Tables 11.7 and 11.8 for the calculation of the respective screening functions], we can obtain the scattering amplitudes f_H, g_H, f_O, and g_O for both H and O (as in the previous exercise). The differential elastic scattering cross-section can thus be calculated by using Eq. (9.131). For the case of H_2O, it can be written

Table 11.7: γ_i ($i = 1, \ldots, 10$) parameters of H and O according to Cox and Bonham [7].

Z	γ_1	γ_2	γ_3	γ_4	γ_5	γ_6	γ_7	γ_8	γ_9	γ_{10}
1	0.0524	5.0360	−4.0876	0.0000	0.0000	0.0000	0.0000	0.0000	0.0000	0.0000
8	1.3017	−0.1670	2.6221	1.5881	−2.8644	−1.4804	0.0000	0.0000	0.0000	0.0000

Table 11.8: λ_i ($i = 1, \ldots, 10$) parameters of H and O according to Cox and Bonham [7].

Z	λ_1	λ_2	λ_3	λ_4	λ_5	λ_6	λ_7	λ_8	λ_9	λ_{10}
1	1.9986	1.8954	2.1161	0.0000	0.0000	0.0000	0.0000	0.0000	0.0000	0.0000
8	2.2491	19.5541	6.9101	10.7798	6.0560	9.9776	0.0000	0.0000	0.0000	0.0000

as

$$\left(\frac{d\sigma}{d\Omega}\right)_{H_2O} = 2\left(\frac{d\sigma}{d\Omega}\right)_{H} + \left(\frac{d\sigma}{d\Omega}\right)_{O}$$
$$+ 2\frac{\sin qr_{OH}}{qr_{OH}}[f_O f_H^* + f_H f_O^* + g_O g_H^* + g_H g_O^*]$$
$$+ 2\frac{\sin qr_{HH}}{qr_{HH}}(|f_H|^2 + |g_H|^2), \tag{11.68}$$

where $r_{OH} = 0.9572\,\text{Å}$ and $r_{HH} = 1.514\,\text{Å}$. Please note that the use of the additivity approximation:

$$\left(\frac{d\sigma_{el}}{d\Omega}\right)_{H_2O} = 2\left(\frac{d\sigma_{el}}{d\Omega}\right)_{H} + \left(\frac{d\sigma_{el}}{d\Omega}\right)_{O}, \tag{11.69}$$

provides results in very good agreement (for this electron energy) with the use of Eq (11.68), the differences becoming appreciable for only small angles of scattering [see Fig. 11.19].

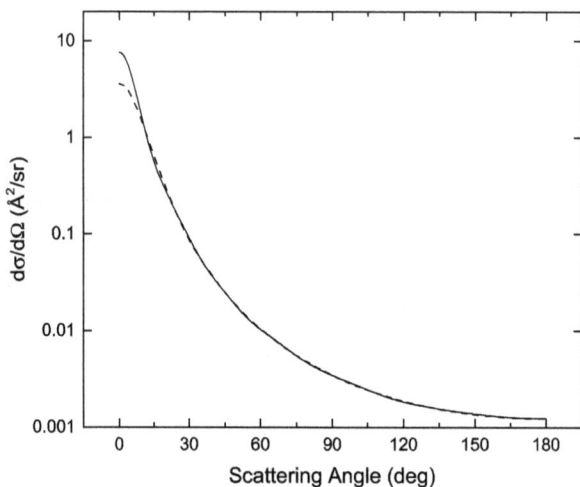

Figure 11.19: The differential elastic scattering cross-section of 1,000 eV electrons in H_2O calculated using the Dirac equation (relativistic partial wave expansion method). Cox and Bonham screening function [7] was used to calculate the electrostatic potential. Solid line: Eq. (11.68). Dashed line: Eq. (11.69).

11.22 Exercise 22

Calculate the differential elastic scattering cross-section for 1,000 eV electrons in H_2O using the Mott theory. Compute the atomic potential using the Salvat et al. screening function [24] and including the exchange effects according to the Furness and Mc-Carthy formula [13].

Solution

The differential elastic scattering cross-section, calculated by using Eq. (11.68), is presented in Fig. 11.20. Please note that the parameters for the calculation of the screening function can be found in Table 11.9 and that the exchange effect can be taken into account by using the Furness and McCarthy formula [13]:

$$V_{ex} = \frac{1}{2}(E - V) - \frac{1}{2}\sqrt{(E - V)^2 + 4\pi\, a_0\, e^4 \rho},$$ (11.70)

where E is the electron kinetic energy, $V = V(r)$ the electrostatic scalar potential, e the electron charge, a_0 the Bohr radius, and ρ the atomic electron density (obtained from Poisson's equation).

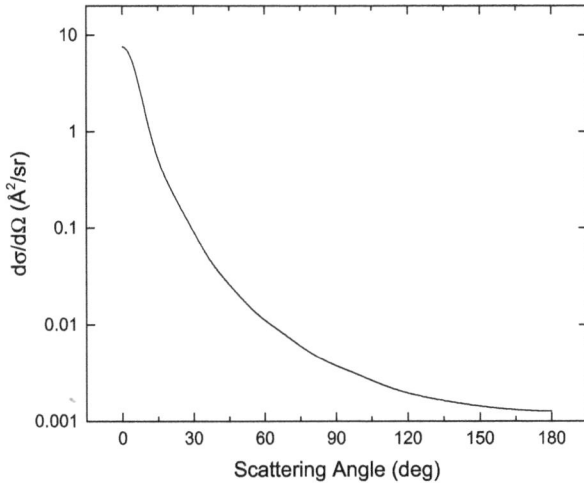

Figure 11.20: The differential elastic scattering cross-section of 1,000 eV electrons in H_2O calculated using the Dirac equation (relativistic partial wave expansion method). Salvat et al. screening function [24] was used to calculate the electrostatic potential. Furness and McCarthy exchange effect [13] was included in the calculation of the atomic potential.

Table 11.9: γ_i and λ_i ($i = 1, 2, 3$) parameters of H and O according to Salvat et al. [24].

Z	γ_1	γ_2	γ_3	λ_1	λ_2	λ_3
1	−184.39	185.39	0.0000	2.0027	1.9973	0.0000
8	0.0625	0.9375	0.0000	14.823	2.0403	0.0000

11.23 Exercise 23

Calculate the Sherman function $S(\vartheta)$ of 1,000 eV, 1,500 eV, and 2,000 eV electrons impinging on Au atoms. Compute the atomic potential using the Salvat et al. screening function [24].

Solution

Once known, the scattering amplitudes $f(\vartheta)$ and $g(\vartheta)$ (calculated using the relativistic partial wave expansion method), the Sherman function can be computed by using Eq. (9.62). It is represented in Fig. 11.21 for 1,000 eV, 1,500 eV, and 2,000 eV electrons impinging on gold atoms.

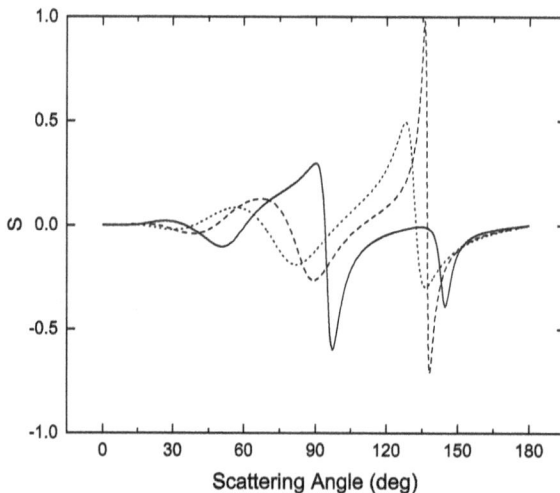

Figure 11.21: The Sherman function $S(\vartheta)$ of 1,000 eV (solid line), 1,500 eV (dashed line), and 2,000 eV (dotted line) electrons impinging on Au atoms calculated using the Dirac equation (relativistic partial wave expansion method). The Salvat et al. screening function [24] was used to calculate the electrostatic potential.

11.24 Exercise 24

Calculate the functions $T(\vartheta)$ and $U(\vartheta)$ of 1,500 eV electrons impinging on Au atoms. Compute the atomic potential using the Salvat et al. screening function [24].

Solution

Once known, the scattering amplitudes $f(\vartheta)$ and $g(\vartheta)$ (calculated using the relativistic partial wave expansion method), the functions $T(\vartheta)$ and $U(\vartheta)$ can be computed by using Eqs. (10.84) and (10.85). They are represented in Fig. 11.22 for 1,500 eV electrons impinging on gold atoms.

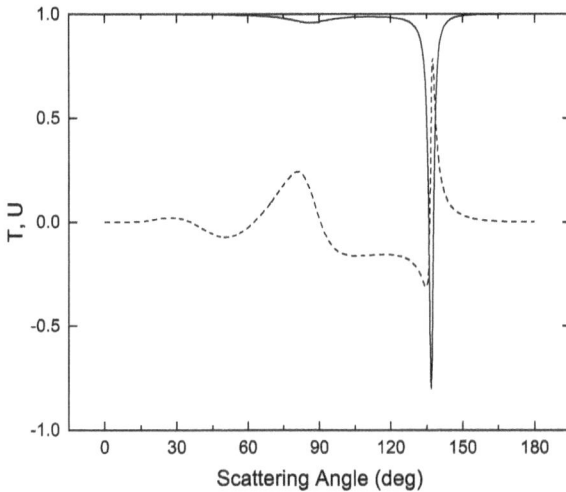

Figure 11.22: $T(\vartheta)$ (solid line) and $U(\vartheta)$ (dashed line) functions of 1,500 eV electrons impinging on Au atoms calculated using the Dirac equation (relativistic partial wave expansion method). The Salvat et al. screening function [24] was used to calculate the electrostatic potential.

11.25 Exercise 25

Calculate the cumulative probability $P(\theta)$ for 1,500 eV electrons impinging on Au atoms. Use the relativistic partial wave expansion method. Compute the atomic potential using the Salvat et al. screening function [24].

Solution

The cumulative probability $P(\theta)$ is given by

$$P(\theta) = \frac{2\pi \int_0^\theta (d\sigma/d\Omega) \sin \vartheta \, d\vartheta}{2\pi \int_0^\pi (d\sigma/d\Omega) \sin \vartheta \, d\vartheta} = \frac{2\pi}{\sigma} \int_0^\theta \frac{d\sigma}{d\Omega} \sin \vartheta \, d\vartheta, \tag{11.71}$$

where σ is the total elastic scattering cross-section:

$$\sigma = 2\pi \int_0^\pi \frac{d\sigma}{d\Omega} \sin \vartheta \, d\vartheta. \tag{11.72}$$

The function $P(\theta)$, represented in Fig. 11.23 for 1,500 eV electrons impinging on gold atoms, was obtained using the Bode quadrature rule to calculate the integrals.

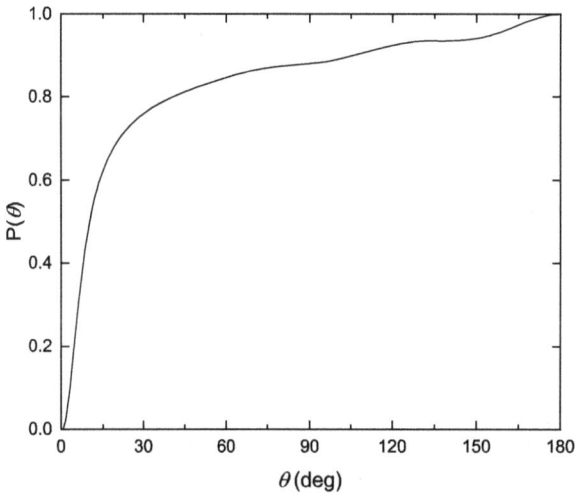

Figure 11.23: Cumulative probability $P(\theta)$ for 1,500 eV electrons impinging on Au atoms calculated using the Dirac equation (relativistic partial wave expansion method). The Salvat et al. screening function [24] was used to calculate the electrostatic potential.

11.26 Exercise 26

Calculate the transport cross-section of electrons impinging on Al atoms in the energy range from 100 eV to 100 keV. Use the relativistic partial wave expansion method. Compute the atomic potential using the Cox and Bonham screening function [7]. Include the exchange effect by using the Furness and McCarthy formula [13].

Solution

The transport cross-section is given by

$$\sigma_{tr} = 2\pi \int_0^\pi (1 - \cos\vartheta) \frac{d\sigma}{d\Omega} \sin\vartheta \, d\vartheta. \tag{11.73}$$

The transport cross-sections of electrons of 100 eV, 1,000 eV, 10,000 eV, and 100,000 eV impinging on Al were obtained using the Bode quadrature rule and are presented in Table 11.10.

Table 11.10: Transport cross-sections σ_{tr} of electrons (kinetic energy from 100 eV to 100 keV) imping-ing on Al. Cox and Bonham screening function [7] was used to calculate the electrostatic potential. The Furness and McCarthy exchange effect [13] was included in the calculation of the atomic poten-tial.

Electron energy (eV)	σ_{tr} (Å^2)
10^2	2.11
10^3	1.20×10^{-1}
10^4	2.68×10^{-3}
10^5	4.56×10^{-5}

11.27 Exercise 27

Calculate the transport cross-section of positrons impinging on Al atoms in the en-ergy range from 100 eV to 100 keV. Use the relativistic partial wave expansion method. Compute the atomic potential using the Cox and Bonham screening function [7].

Solution

Since the impinging particles are positrons, the potential energy can be calculated as

$$V(r) = \frac{Ze^2}{r} \xi(r), \tag{11.74}$$

where $\xi(r)$ is the Cox and Bonham screening function [7]. The transport cross-sections of positrons of 100 eV, 1,000 eV, 10,000 eV, and 100,000 eV impinging on Al were ob-tained using the Bode quadrature rule and are presented in Table 11.11.

Table 11.11: Transport cross-sections σ_{tr} of positrons (kinetic energy from 100 eV to 100 keV) impinging on Al. The Cox and Bonham screening function [7] was used to calculate the electrostatic potential.

Positron energy (eV)	σ_{tr} (Å^2)
10^2	$6.90\ 10^{-1}$
10^3	$6.17\ 10^{-2}$
10^4	$2.24\ 10^{-3}$
10^5	$4.30\ 10^{-5}$

Bibliography

[1] M. Abramowitz and I. A. Stegun (Edited by), *Handbook of Mathematical Functions* (Dover Publications, 1972).

[2] H. A. Bethe and R. Jackiw, *Intermediate Quantum Mechanics* (Benjamin, 1968).

[3] P. J. Bunyan and J. L. Schonfelder, *Polarization by Mercury of 100 to 2000 eV Electrons* (Proc. Phys. Soc., Vol. 85, pp. 455–463, 1965).

[4] R. L. Burden and J. D. Faires, *Numerical Analysis* (PWS Publishers, 1985).

[5] P. G. Burke and C. J. Joachain, *Theory of Electron–Atom Collisions* (Plenum Press, 1995).

[6] R. V. Churchill, *Fourier Series and Boundary Value Problems* (McGraw-Hill, 1963).

[7] H. L. Cox Jr. and R. A. Bonham, *Elastic Electron Scattering Amplitudes for Neutral Atoms Calculated Using the Partial Wave Method at 10, 40, 70, and 100 kV for Z = 1 to Z = 54* (J. Chem. Phys., Vol. 47, pp. 2599–2608, 1967).

[8] M. Dapor, *Electron-Beam Interactions with Solids. Application of the Monta Carlo Method to Electron Scattering Problems* (Springer, 2003).

[9] M. Dapor, *Relatività e Meccanica Quantistica Relativistica*, Edited by G. Introzzi (Carocci, 2011).

[10] M. Dapor, *Transport of Energetic Electrons in Solids. Computer Simulation with Applications to Materials Analysis and Characterization* (Springer, 2020).

[11] J. E. G. Farina, *Scattering of a Particle by a Centre of Force*, in *The International Encyclopedia of Physical Chemistry and Chemical Physics* (Vol. 4, Pergamonn, 1973).

[12] L. L. Foldy, *Relativistic Wave Equations*, in *Quantum Theory III. Radiation and High Energy Physics*, Edited by R. Bates (Academic Press, 1962).

[13] J. B. Furness and I. E. McCarthy, *Semiphenomenological Optical Model for Electron Scattering on Atoms* (J. Phys. B, At. Mol. Phys., Vol. 6, pp. 2280–2291, 1973).

[14] J. Kessler, *Polarized Electrons* (Springer, 1985).

[15] S. E. Koonin and D. C. Meredith, *Computational Physics* (Addison-Wesley, 1990).

[16] L. D. Landau and E. M. Lifsits, *The Classical Theory of Fields. Vol. 2* (Addison-Wesley, 1951).

[17] S.-R. Lin, N., Sherman, and J. K. Percus, *Elastic Scattering of Relativistic Electrons by Screened Atomic Nuclei* (Nucl. Phys., Vol. 45, pp. 492–504, 1963).

[18] A. Messiah, *Quantum Mechanics I and II* (North Holland, 1961).

[19] N. F. Mott, *The Scattering of Fast Electrons by Atomic Nuclei* (Proc. R. Soc. Lond. A, Vol. 124, pp. 425–442, 1929).

[20] W. H. Press, S. A. Teukolsky, W. T. Vetterling, and B. P. Flannery, *Numerical Recipes* (Cambridge University Press, 2007).

[21] M. Reiher and A. Wolf, *Relativistic Quantum Chemistry. The Fundamental Theory of Molecular Science* (Wiley, 2015).

[22] L. S. Rodberg and R. M. Thaler, *Introduction to the Quantum Theory of Scattering* (Academic Press, 1967).

[23] F. Salvat, A. Jablonski, and C. J. Powell, *ELSEPA – Dirac partial-wave calculation of elastic scattering of electrons and positrons by atoms, positive ions and molecules* (Computer Physics Communications, Vol. 165, pp. 157–190, 2005).

[24] F. Salvat, J. D. Martinez, R. Mayol, and J. Parellada, *Analytical Dirac–Hartree–Fock–Slater Screening Function for Atoms (Z = 1 – 92)* (Phys. Rev. A, Vol. 36, pp. 467–474, 1987).

[25] F. Salvat and R. Mayol, *Elastic Scattering of Electrons and Positrons by Atoms. Schrödinger and Dirac Partial Wave Analysis* (Comput. Phys. Commun., Vol. 74, pp. 358–374, 1993).

[26] F. Schwabl, *Quantum Mechanics* (Springer, 1992).

[27] F. Schwabl, *Advanced Quantum Mechanics* (Springer, 1997).

[28] P. Strange, *Relativistic Quantum Mechanics* (Cambridge University Press, 1998).

[29] L. Susskind and A. Friedman, *Special Relativity and Classical Field Theory. The Theoretical Minimum* (Penguin Books, 2018).

https://doi.org/10.1515/9783110675375-012

[30] S. Weimberg, in *Equilibrio Perfetto* (il Saggiatore, 2005, pp. 345–351).

[31] S. Weimberg, *Lectures on Quantum Mechanics* (Cambridge University Press, 2nd Edition, 2015).

[32] S. Weimberg, *Foundations of Modern Physics* (Cambridge University Press, 2021).

[33] F. Wilczek, in *Equilibrio Perfetto* (il Saggiatore, 2005, pp. 192–227).

Index

Adams–Bashforth method 3, 8, 9
Angular momentum 12, 41, 46, 97
Associated Legendre functions 12–14, 149, 150
Atomic potential 28, 51–53, 57, 59, 111,
    157–159, 166, 168, 169, 171–175

Bessel functions 12, 14, 15, 31, 32, 115, 124,
    154–158
Bode quadrature rule 3, 5, 174
Boost 68

Central field 97, 104, 113
Central potential 37, 39
Computational physics 3
Confluent hypergeometric function 12, 17, 18,
    107
Continuity equation 24, 25, 78
Contraction 70
Contravariant four-vectors 69, 70
Covariant four-vectors 69–71
Cross-section 21, 25, 27, 36, 39, 40, 51, 112,
    118, 120, 121, 125, 136, 140, 141, 157–159,
    168–171, 174–176
Cumulative probability 173, 174

d'Alembert equation 22
d'Alembert operator 72, 74–76
Density matrix 130–132, 134, 137
Differential elastic scattering cross-section 25,
    27, 36, 51, 112, 120, 125, 136, 140, 141,
    157–159, 168–171
Dirac delta distribution 13
Dirac equation 65, 80, 81, 83, 86–88, 90, 91,
    94, 95, 97, 102, 103, 109, 112, 113, 169–174
Dirac matrices 86, 89
Dirac notation 42, 130
Dirac theory 88, 91, 95–97, 104, 108, 130
Dirac–Hartree–Fock self-consistent method
    109, 110
Dirac–Hartree–Fock–Slater self-consistent
    method 109

Effective Dirac potential 114
Eigenfunctions 12, 16, 54
Eigenspace 46
Eigenvalues 16, 29, 42, 44, 46, 51, 54, 55, 81,
    92, 102

Eigenvectors 42, 44, 46, 47, 130
Einstein convention of summing over repeated
    indices 69, 70, 88
Elastic scattering 37, 40
Electromagnetic four-tensor 72, 90
Electron density 53, 56, 57
Electron–molecule elastic scattering 128
Energy levels 97, 108, 109
Euler method 3, 8
Exchange effects 171, 175

Fermi energy 56
Fermi gas 53, 56
First Born approximation 37–39
Four-potential 72, 73, 87, 90
Four-vectors 67, 69–72, 74
Fourth-order Runge–Kutta method 9, 125, 167

Gauss law 56
Gaussian quadrature rule 3, 5
Green function 12, 18, 37

Hartree–Fock self-consistent method 51, 59, 61,
    62

Internal multiplication 70
Invariance of the interval 67
Irregular spherical Bessel functions 12, 15, 115,
    124, 154, 156, 158

Klein–Gordon equation 65, 73–79, 90
Kronecker symbol 13, 70

Laplacian operator 21, 29
Legendre polynomials 12–14, 28–33, 147–149,
    158
Linear second-order differential equations 10
Lorentz transformation 66–69

Manifest covariance 70, 86, 87
Metric tensor 69, 70, 89
Mixed states 130
Monte Carlo method 3, 6, 7

Natural system of units 65
Numerical analysis 3
Numerical differentiation 3, 155

Numerical integration 3, 121, 125
Numerical quadrature 3, 4
Numerov algorithm 3, 10, 11, 51, 157, 158

Operators 16, 18, 21–23, 29, 37, 41, 42, 46, 49,
    71, 74–76, 81, 86, 88, 89, 92, 95, 97–100,
    102, 130–132
Optical theorem 36
Ordinary differential equations 3, 8
Orthonormal eigenvectors 130

Partial wave expansion method 21, 157, 158
Partial waves 21, 28, 31, 112, 157, 158, 169–175
Pauli matrices 48–50, 81, 85, 94, 99, 119, 132
Phase shifts 30, 31, 51, 52, 112, 115, 122–125,
    127, 158, 166–169
Plane waves 21, 25
Poincaré transformation 69
Proper time 67
Pure states 130, 131

Quantum-relativistic theory 130

Radial equation 28, 30, 31, 51, 52, 102, 104, 157
Radial-momentum operator 100
Recursion relations 12, 14, 16, 148, 149, 151,
    152, 156
Regular spherical Bessel functions 12, 15, 124
Relativistic partial wave expansion method 112,
    169–175
Runge–Kutta method 3, 9
Rutherford elastic scattering cross-section 39
Rutherford formula 40

Scattering amplitudes 26, 35, 36, 112, 115, 118,
    158, 168, 169
Schrödinger equation 14, 23, 25–28, 53, 65,
    73–75, 77, 78, 109, 157, 158
Screening function 51, 53, 59, 109, 110, 126,
    159–162, 165–176
Sherman function 119, 138, 140, 142, 172
Simpson quadrature rule 3, 5
Special functions 12
Spectrum of angular momentum 41
Spherical Bessel functions 12, 15
Spherical harmonics 12, 16, 17, 115
Spherical Neumann functions 15
Spherical waves 21, 26
Spin 41, 46, 47, 56, 60, 62, 65, 80, 85, 86,
    88–92, 94–96, 99, 112, 117–119, 130, 132
Spin down 93, 104, 108, 112–114, 118, 135
Spin four-tensor 90
Spin operator 92, 95
Spin up 93, 103, 104, 108, 112–114, 117, 118,
    134, 135
Spin-polarization 120, 128, 130, 132, 134, 137,
    140
Spinors 83, 91, 93–95, 103, 112, 119

Tensors 67, 69, 70, 72, 89, 90
Thomas–Fermi atom 51, 53, 56, 57, 59
Total elastic scattering cross-section 21, 36, 121
Transport elastic scattering cross-section 121,
    174–176
Trapezoidal quadrature rule 3, 4

Wentzel-like atomic potential 39, 40

www.ingramcontent.com/pod-product-compliance
Lightning Source LLC
Chambersburg PA
CBHW081526220326
41598CB00036B/6341